T0338590

A Computational Introduction to Quantum Physics

This concise textbook introduces an innovative computational approach to quantum mechanics. Over the course of this engaging and informal book, students are encouraged to take an active role in learning key concepts by working through practical exercises. By equipping readers with some basic methodology and a toolbox of scientific computing methods, they can use code to simulate and directly visualize how quantum particles behave. The important foundational elements of the wave function and the Schrödinger equation are first introduced, then the text gradually builds up to advanced topics including relativistic, open, and non-Hermitian quantum physics.

This book assumes familiarity with basic mathematics and numerical methods, and can be used to support a two-semester advanced undergraduate course. Source code and solutions for every book exercise involving numerical implementation are provided in Python and MATLAB®, along with supplementary data. Additional problems are provided online for instructor use with locked solutions.

Sølve Selstø is a professor of Physics at Oslo Metropolitan University, Norway. His research focuses on computational and theoretical quantum physics. He teaches both physics and mathematics, and believes that understanding within science and mathematics predominantly emerges from working with and discussing real-world problems. Selstø has previously written an introductory textbook on numerical methods, which he believes are valuable for both their utility and their ability to make theory tangible.

'This book provides a well-written and carefully structured course on computational quantum mechanics, and its many exercises, with provided solutions, facilitate active learning both of physics and of numerical computing . . . This book will be very useful for undergraduate and graduate students interested in adopting modern methodologies for their learning by applying computational resources to intermediate to advanced quantum mechanical concepts.'
Christian Hill, author of *Learning Scientific Programming with Python* (2nd ed., 2020) and *Python for Chemists* (2023)

'Selstø's book is an outstanding introduction to the world of quantum mechanics from a practical and computational, yet rigorous and comprehensive, perspective. For a student, the book is highly engaging, approachable, and it provides an exceptional pool of pedagogically guided exercises. I would highly recommend this book for teachers and students, as both a coursebook and self-study material.'
Esa Räsänen, Tampere University

'This book takes the reader from the basics of quantum theory to the core of a range of topics that define the research front of quantum-based sciences. It is an easy-to-read, modern text, with illustrative examples and problems of an analytical and numerical nature. I would recommend the book for any intermediate quantum mechanics course!'
Jan Petter Hansen, University of Bergen

'This book presents, delightfully, a large number of practical tasks on both foundational topics and modern applications, with code provided online. More than a collection of exercises, yet less than a complete monograph, this feels like a workbook, in that the text screams out to be worked through . . . Students of quantum physics who wish to see what the abstract formalism implies in practice are in for a treat.'
Alexandros Gezerlis, University of Guelph

A Computational Introduction to Quantum Physics

Sølve Selstø
Oslo Metropolitan University

Shaftesbury Road, Cambridge CB2 8EA, United Kingdom

One Liberty Plaza, 20th Floor, New York, NY 10006, USA

477 Williamstown Road, Port Melbourne, VIC 3207, Australia

314–321, 3rd Floor, Plot 3, Splendor Forum, Jasola District Centre,
New Delhi – 110025, India

103 Penang Road, #05–06/07, Visioncrest Commercial, Singapore 238467

Cambridge University Press is part of Cambridge University Press & Assessment,
a department of the University of Cambridge.

We share the University's mission to contribute to society through the pursuit of
education, learning and research at the highest international levels of excellence.

www.cambridge.org
Information on this title: www.cambridge.org/highereducation/isbn/9781009389631

DOI: 10.1017/9781009389594

First published 2024

A catalogue record for this publication is available from the British Library

Library of Congress Cataloging-in-Publication Data
Names: Selstø, Sølve, author.
Title: A computational introduction to quantum physics / Sølve Selstø.
Description: Cambridge, United Kingdom ; New York, NY : Cambridge
University Press, 2024. | Includes bibliographical references and index.
Identifiers: LCCN 2023051947 | ISBN 9781009389631 (hardback) | ISBN
9781009389594 (ebook)
Subjects: LCSH: Quantum theory – Mathematics – Textbooks. | Quantum
theory – Data processing – Textbooks.
Classification: LCC QC174.17.M35 S45 2024 | DDC
530.12078/5–dc23/eng/20231212
LC record available at https://lccn.loc.gov/2023051947

ISBN 978-1-009-38963-1 Hardback

Additional resources for this publication at www.cambridge.org/Selsto

To Anne Birte, Halvard and Ingeborg

Contents

Preface page xi
Acknowledgements xiii

1 The Wave Function 1
1.1 Quantum Physics: the Early Years 1
1.2 How Different Is Quantum Physics? 4
1.3 The Wave Function and the Curse of Dimensionality 5
1.4 Interpretation of the Wave Function 7
1.4.1 Exercise: Normalization 7
1.4.2 Exercise: Position Expectation Values 9
1.4.3 Exercise: Is the Particle in This Interval? 10
1.5 The Momentum Operator 10
1.5.1 Exercise: Momentum Expectation Values 11
1.6 Some Simplifying Notation 12
1.6.1 Exercise: Standard Deviations 12
1.6.2 Exercise: The Inner Product 13
1.6.3 Exercise: Hermicity 13
1.6.4 Exercise: The Hydrogen Atom and Atomic Units 14

2 The Schrödinger Equation 16
2.1 Numerical Discretization 17
2.1.1 Exercise: Normalizing the Discretized Wave Function 18
2.2 Kinetic Energy Numerically 19
2.2.1 Exercise: The Kinetic Energy Operator as a Matrix 19
2.2.2 Exercise: Expectation Values as Matrix Multiplications 20
2.3 Dynamics Without Explicit Time Dependence in the Hamiltonian 21
2.3.1 Exercise: The Formal Solution 22
2.3.2 Exercise: Wave Propagation 23
2.3.3 Exercise: Interference 24
2.3.4 Exercise: Expectation Values and Uncertainties 25
2.4 Scattering and Tunnelling 26
2.4.1 Exercise: Scattering on a Barrier 27
2.4.2 Exercise: The Dynamics of a Classical Particle 28
2.4.3 Exercise: Tunnelling 29
2.4.4 Exercise: Tunnelling Again 30

2.5 Stationary Solutions 31
2.5.1 Exercise: Enter the Eigenvalue Problem 31
2.6 Eigenstates, Measurements and Commutators 32
2.6.1 Exercise: The Standard Deviation of an Eigenstate 32
2.6.2 Exercise: Commuting with the Hamiltonian 34

3 The Time-Independent Schrödinger Equation 37
3.1 Quantization 37
3.1.1 Exercise: Bound States of a Rectangular Well 37
3.1.2 Exercise: Solution by Direct Numerical Diagonalization 39
3.1.3 Exercise: The Bohr Formula 41
3.1.4 Exercise: The Harmonic Oscillator 42
3.2 A Glimpse at Periodic Potentials 43
3.2.1 Exercise: Eigenenergies for Periodic Potentials – Bands 43
3.3 The Spectral Theorem 46
3.3.1 Exercise: Trivial Time Evolution 47
3.3.2 Exercise: Glauber States 47
3.4 Finding the Ground State 49
3.4.1 Exercise: The Variational Principle and Imaginary Time 50
3.4.2 Exercise: One-Variable Minimization 52
3.4.3 Exercise: Imaginary Time Propagation 52
3.4.4 Exercise: Two-Variable Minimization, Gradient Descent 53
3.4.5 Exercise: Variational Calculation in Two Dimensions 55

4 Quantum Physics with Several Particles – and Spin 57
4.1 Identical Particles and Spin 57
4.1.1 Exercise: Exchange Symmetry 58
4.1.2 Exercise: On Top of Each Other? 61
4.2 Entanglement 63
4.2.1 Exercise: Entangled or Not Entangled 64
4.3 The Pauli Matrices 66
4.3.1 Exercise: Some Characteristics of the Pauli Matrices 67
4.3.2 Exercise: The Eigenstates of the Pauli Matrices 68
4.4 Slater Determinants, Permanents and Energy Estimates 68
4.4.1 Exercise: Repeated States in a Slater Determinant 69
4.4.2 Exercise: The Singlet and Triplet Revisited 70
4.4.3 Exercise: Variational Calculations with Two Particles 70
4.4.4 Exercise: Self-Consistent Field 74

5 Quantum Physics with Explicit Time Dependence 78
5.1 Dynamics with Spin-1/2 Particles 78
5.1.1 Exercise: Dynamics with a Constant Magnetic Field 80
5.1.2 Exercise: Dynamics with an Oscillating Magnetic Field 81
5.1.3 Exercise: The Rotating Wave Approximation 82
5.1.4 Exercise: Spin Dynamics with Two Spin-1/2 Particles 84

5.2 Propagation 86
5.2.1 Exercise: Magnus Propagators of First and Second Order 86
5.2.2 Exercise: Split Operators 87
5.2.3 Exercise: Photoionization 88
5.3 Spectral Methods 90
5.3.1 Exercise: The Matrix Version 91
5.3.2 Exercise: Dynamics Using a Spectral Basis 92
5.3.3 Exercise: Momentum Distributions 93
5.4 'Dynamics' with Two Particles 95
5.4.1 Exercise: Symmetries of the Ψ-Matrix 95
5.4.2 Exercise: The Two-Particle Ground State 96
5.5 The Adiabatic Theorem 97
5.5.1 Exercise: Adiabatic Evolution 97
5.5.2 Exercise: A Slowly Varying Potential 99

6 Quantum Technology and Applications 101
6.1 Scanning Tunnelling Microscopy 101
6.1.1 Exercise: Tunnelling Revisited 102
6.1.2 Exercise: The Shape of a Surface 105
6.2 Spectroscopy 107
6.2.1 Exercise: Emission Spectra of Hydrogen 108
6.2.2 Exercise: The Helium Spectrum 108
6.3 Nuclear Magnetic Resonance 110
6.3.1 Exercise: Spin Flipping – On and Off Resonance 110
6.4 The Building Blocks of Quantum Computing 114
6.4.1 Exercise: A Blessing of Dimensionality 114
6.4.2 Exercise: The Qubit 116
6.4.3 Exercise: Quantum Gates and Propagators 117
6.4.4 Exercise: Quantum Gates are Unitary 117
6.4.5 Exercise: Pauli Rotations 118
6.4.6 Exercise: CNOT and SWAP 119
6.4.7 Exercise: Prepare Bell 120
6.5 Quantum Protocols and Quantum Advantage 121
6.5.1 Exercise: Superdense Coding 123
6.5.2 Exercise: Quantum Key Distribution 123
6.6 Adiabatic Quantum Computing 126
6.6.1 Exercise: Quantum Minimization 127

7 Beyond the Schrödinger Equation 130
7.1 Relativistic Quantum Physics 130
7.1.1 Exercise: Relativistic Kinetic Energy 130
7.1.2 Exercise: Arriving at the Dirac Equation 132
7.1.3 Exercise: Eigenstates of the Dirac Hamiltonian 134
7.1.4 Exercise: The Non-relativistic Limit of the Dirac Equation 137

7.2 Open Quantum Systems and Master Equations 138
7.2.1 Exercise: The von Neumann Equation 139
7.2.2 Exercise: Pure States, Entanglement and Purity 140
7.2.3 Exercise: Two Spin-1/2 Particles Again 141
7.2.4 Exercise: Preserving Trace and Positivity 143
7.2.5 Exercise: A Decaying Quantum Bit 145
7.2.6 Exercise: Capturing a Particle 148

8 Non-Hermitian Quantum Physics 152
8.1 Absorbing Boundary Conditions 152
8.1.1 Exercise: Decreasing Norm 153
8.1.2 Exercise: Photoionization with an Absorber 153
8.1.3 Exercise: Scattering with an Absorber 154
8.2 Resonances 155
8.2.1 Exercise: Scattering off a Double Well 155
8.2.2 Exercise: Outgoing Boundary Conditions 157
8.2.3 Exercise: The Lifetime of a Resonance 160
8.2.4 Exercise: Doubly Excited States 162

9 Some Perspective 165
9.1 But What Does It Mean? 166
9.2 Quantum Strangeness 166
9.3 What We Didn't Talk About 167

References 171
Figure Credits 174
Index 176

Preface

This book is meant to give you a bit of a feel for quantum physics, which is the mathematical description of the how the micro world behaves. Since all matter is built up of small particles belonging to the quantum world, this is a theory of rather fundamental importance. It is only fair that we spend some time learning about it. I believe the best way to do so is by exploring it for ourselves, rather than just reading about it or having someone telling us about it.

That is what this book invites you to do. Most of the book consists of exercises. These exercises are not there in order to supplement the text. It is rather the opposite. What you will learn about quantum physics, you will learn from working out these examples. It makes no sense to disregard them and resort to the text alone. My hope is that, from working out these exercises, by both numerical and analytical means, you will learn that the quantum world is surprisingly interesting – quirky and beautiful.

It may be an advantage to have a certain familiarity with classical physics – having a feel for concepts such as kinetic and potential energy, momentum and velocity. Knowing some basic concepts from *statistics* may come in handy as well. However, *mathematics* is more important here. In order to *practice* quantum physics, the proper language and framework is that of mathematics and numerics. You must be familiar with basic calculus and linear algebra – both numerically and analytically. Topics such as differentiation, Taylor expansions, integration, differential equations, vectors, matrix operations and eigenvalues should be quite familiar. The quantities to be calculated and visualized will predominantly be obtained by numerical means. Thus, it is crucial that you have some experience in doing so. This, in turn, requires familiarity with a relevant software environment or programming language, be it Python, MATLAB®/Octave, Julia, Java, C/C++ or anything similar.

In solving the exercises, you are not entirely on your own. In addition to, possibly, having fellow students and teachers, you will also have access to a solutions manual and a GitHub repository with source code. Please use these resources cautiously. They may be helpful in finding the way forward when you struggle or when you want to compare approaches. But they are not that helpful if you resort to these suggestions without having made a proper effort yourself first.

In addition to these solutions, the online material also features colour illustrations and relevant data.

This book strives to illuminate quantum phenomena using examples that are as simple and clear cut as possible – while not being directly banal. As a consequence, many exercises are confined to a one-dimensional world. While this, of course, differs from the actual dimensionality of the world, it still allows the introduction of non-intuitive quantum phenomena without excessive technical complications.

Now, let's play!

Acknowledgements

I have had the pleasure of interfering constructively with several mentors and collaborators over the years. Some deserve particular mention. In alphabetical order: Sergiy Denisov, Morten Førre, Michael Genkin, Jan Petter Hansen, Tor Kjellsson Lindblom, Ladislav Kocbach, Simen Kvaal and Eva Lindroth. The insights I have gained from you over the years have certainly affected the outcome of this book.

Sergiy Denysov, Morten Førre and Eva Lindroth have also taken active parts in providing constructive feedback on the manuscript – along with Joakim Bergli, André Laestadius, Tulpesh Patel and Konrad Tywoniuk. Specific input from Knut Børve, Stefanos Carlström and Trygve Helgaker is also gratefully acknowledged.

In trying to find illustrations to use, I have met an impressive amount of good will. Several people and enterprises have generously shared their graphical material – and in some cases even produced images. In this regard, I am particularly indebted to Reiner Blatt and Jonas Østhassel.

Several of my students at the Oslo Metropolitan University have been exposed to early versions of this manuscript. Their experiences led to suggested adjustments which, hopefully, have improved the quality and accessibility of the book. Oslo Metropolitan University also deserves my gratitude for providing a stimulating scientific environment – and for flexibility and goodwill when it came to spending time writing this book.

Finally, I am eternally grateful to my first physics teacher, Helge Dahle, for igniting the passion for physics and science in the first place.

1 The Wave Function

The physical world we see around us seems to follow the well-known laws of Isaac Newton. For a point object of mass m, his second law may be written

$$m \frac{\mathrm{d}^2}{\mathrm{d}t^2}\mathbf{r}(t) = \mathbf{F}(\mathbf{r}, t), \tag{1.1}$$

where $\mathbf{r}(t)$ is the position of the object at time t and $\mathbf{F}$ is the total force acting on the object. Often, this force is the gradient of some potential V, which may be time dependent:[1]

$$\mathbf{F}(\mathbf{r}, t) = -\nabla V(\mathbf{r}, t). \tag{1.2}$$

This potential could, for instance, be the gravitational attraction of a planet, in which case we would have

$$V(r) = -k\,\frac{Mm}{r}, \tag{1.3}$$

where k is a constant, M is the mass of the planet and r is the distance between the planet and our object.

This is Newton's law of gravitation. Early in the twentieth century it became clear that this law, which had been immensely successful until then, was not entirely accurate. With his *general theory of relativity*, Albert Einstein suggested that Eq. (1.3) would not describe massive gravitational objects correctly. In time, his theory gained support from observations, and it is now considered the proper description of gravitation.

More or less at the same time, it also became clear that Newtonian mechanics just wouldn't do at the other end of the length scale either – for objects belonging to the micro cosmos. For several reasons, another description of microscopic objects such as molecules, atoms and elementary particles was called for.

1.1 Quantum Physics: the Early Years

Once again, Albert Einstein had quite a lot to do with resolving these issues. But where the *theories of relativity*, both the *special* one and the *general* one, to a large extent are

[1] This is the last time we will use the **F**-word in this context; the term simply does not apply in quantum physics, instead we talk about *potentials* and *interactions*.

products of one single brilliant mind, the development of our understanding of the micro world is the beautiful product of several brilliant minds contributing constructively. The story leading up to the birth of *quantum physics* is a truly interesting one. It's worth mentioning a few highlights.

At the time, around of the turn of the nineteenth century, the understanding of light as travelling waves was well established. The famous double-slit experiment of Thomas Young was one of several that had enforced such an understanding (see Fig. 1.1). And the Scot James Clerk Maxwell (Fig. 1.2) had provided a perfectly accurate description of the wave nature of electromagnetic radiation, including the electromagnetic radiation that is visible light.

However, about a century after Young presented his interference experiment, two phenomena related to light were observed that just didn't fit with this wave description. Max Planck, from Germany, resolved one of them and our friend Albert succeeded in explaining the other. Both of them did so by assuming that electromagnetic radiation, *light*, was made up of small energy lumps, or *quanta*. In Planck's case, he managed to understand why the radiation emitted from a so-called *black body* at thermal equilibrium is distributed as it is. In doing so, he imposed a simple but non-intuitive relation between the frequency of the radiation and the energy of each quantum:

$$E_{\text{quant}} = hf, \tag{1.4}$$

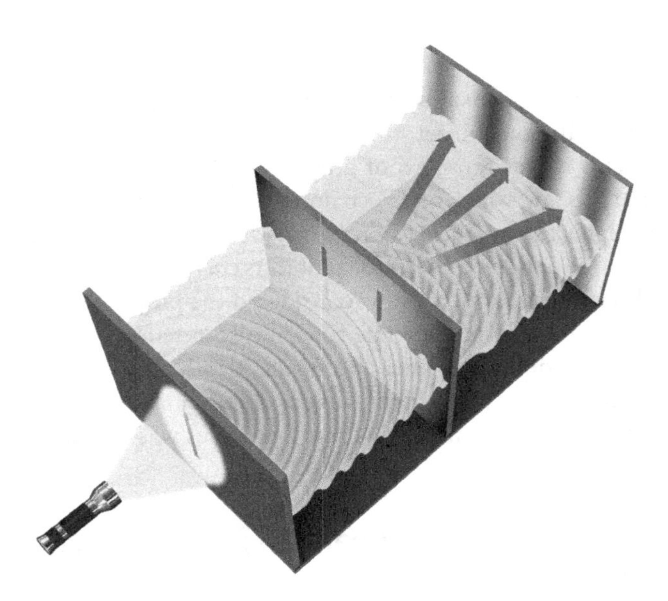

Figure 1.1 In the year 1801, the British scientist Thomas Young, 'the last man who knew everything' [33], showed with his famous double-slit experiment that light *must* be perceived as waves. He sent light of a specific colour, which, in turn, corresponds to a specific wavelength, through a double slit and observed an interference pattern on a screen behind the slits. This pattern emerges from interference between waves originating from each of the two slits; they either reinforce each other or reduce each other, depending on their respective phases where they meet.

Figure 1.2 James Clerk Maxwell as a young and promising student at Trinity College, Cambridge, UK.

where f is the frequency and $h \approx 6.63 \times 10^{-34}$ J s is what is now called the *Planck constant*. Clearly, it's not very large. But it isn't zero either, and this has huge implications.

The same relation was also crucial in Einstein's explanation of the *photoelectric effect*, in which electrons are torn loose from a metal plate by shining light upon it.

These works of Planck and Einstein certainly did not dismiss the idea that light consists of waves. They just showed that, complementary to this concept, we also need to consider light as quanta. We call these quanta *photons*.

The next major conceptual breakthrough is an equally impressive one. At the time, the notion of matter being made up of atoms was finally gaining recognition amongst scientists; it was eventually established that matter indeed consists of particles. But, just as light turned out to be not just waves, the perception of matter as particles also turned out to be an incomplete one. For instance, according to Newton's and Maxwell's equations, an atom couldn't really be stable. It was understood that within the atom a small, negatively charged particle – the *electron* – would orbit a larger positively charged atomic nucleus somehow. However, according to Maxwell's equations, *any* charged particle undergoing acceleration should emit radiation and lose energy. This should also apply to any electron that does not follow a straight line.

A different understanding of the atom was called for.

In his PhD dissertation, the French nobleman Louis de Broglie launched the following thought:[2]

[2] For the sake of clarity: this is not actually a quote.

> Since light turns out to behave not just as waves but also as particles, maybe matter isn't just particles either – maybe matter behaves as waves as well.

And he was right! This idea quickly gained experimental and theoretical support. Electrons were seen to produce interference patterns, just like light had done in the Young experiment. In fact, after a while, his double-slit experiment was repeated – with electrons. We will return to this experiment in Chapter 2.

De Broglie's realization, in turn, called for a mathematical wave description of matter. We needed a scientist to formulate this – just like Maxwell had done for light. We got two: Werner Heisenberg and Erwin Schrödinger. Both their formulations are still applied today. Schrödinger's formulation is, however, the predominant one. And it's his formulation we will use in the following.

In this context, it's worth mentioning that the Dane Niels Bohr used the ideas of de Brogile in order to formulate a 'wave theory' for the atom. This model somehow managed to predict the right energies for the hydrogen atom, and it was an important stepping stone towards our understanding of the atom. However, the 'Bohr atom', which portrays the atom as a miniature solar system, does not provide a very adequate model. It is neither accurate, general nor beautiful enough, and the theory is long since abandoned.

The birth of quantum physics is truly an impressive collaborative effort. The names of several of the *midwives* involved have been mentioned. It is nice to also have faces attached to these names. Very many of them can be seen in the famous photo in Fig. 1.3, which was taken at the fifth Solvay conference, held in Brussels in 1927.

1.2 How Different Is Quantum Physics?

The formulation in which small particles are described in terms of waves is known as *quantum physics*. It's fair to ask in which way and to what extent this description differs from a Newtonian, classical picture – one in which atoms and particles behave as little 'balls' bouncing around. The answer is that quantum physics is *very, very different* from classical physics – in several aspects. This includes phenomena that really couldn't have been foreseen even with the strongest imagination.

In Newtonian mechanics, an object has a well-defined position and velocity at all times. A quantum object doesn't. A wave isn't localized to a single point in space. And when it travels, it typically does so dispersively; it spreads out. The more confined the wave is spatially, the wider is its distribution in velocity – or momentum.[3] If the position x is known, the momentum p is completely undetermined – and *vice versa*. Heisenberg's uncertainty principle states this quite precisely:

$$\sigma_x \, \sigma_p \geq \frac{h}{4\pi}. \tag{1.5}$$

[3] The *momentum* of a particle is the product of its velocity and its mass, $p = mv$.

Figure 1.3 Not just any crowd: 17 of the 29 attendees at the fifth Solvay conference in 1927 were or went on to become Nobel Prize laureates. Marie Skłodowska Curie, front row, third from left, even received two Nobel Prizes. Many of the people mentioned above are here – along with a few other brilliant minds. *Front row, left to right*: Irving Langmuir, Max Planck, Marie Skłodowska Curie, Hendrik Lorentz, Albert Einstein, Paul Langevin, Charles-Eugène Guye, Charles Thomson Rees Wilson, Owen Willans Richardson. *Middle row*: Peter Debye, Martin Knudsen, William Lawrence Bragg, Hendrik Anthony Kramers, Paul Dirac, Arthur Compton, Louis de Broglie, Max Born, Niels Bohr. *Back row*: Auguste Piccard, Émile Henriot, Paul Ehrenfest, Édouard Herzen, Théophile de Donder, Erwin Schrödinger, Jules-Émile Verschaffelt, Wolfgang Pauli, Werner Heisenberg, Ralph Howard Fowler, Léon Brillouin.

Here σ_x is the uncertainty in position and σ_p is the uncertainty in momentum; h is the Planck constant, which we first encountered in Eq. (1.4). In other words, if we want to have information about both position and momentum of a small object, there is a fundamental limit to the accuracy of this information. This is not a limit imposed by practical issues such as the quality of our measurement devices; this seems to be a limit inherent in nature itself.

And, as we will see, it gets stranger.

First, however, we need to say a few words about the *wave function*.

1.3 The Wave Function and the Curse of Dimensionality

Within a quantum system, such as an atom, all information is contained in its *wave function*. This function, in turn, is provided by a set of appropriate initial conditions and Schrödinger's famous equation. We will get back to this equation – for sure.

For a system consisting of, say, N particles, the wave function is a complex function Ψ depending on the coordinates of all particles and, parametrically, on time:

$$\Psi = \Psi(\mathbf{r}_1, \mathbf{r}_2, \ldots, \mathbf{r}_N; t). \tag{1.6}$$

It's *complex* in a double sense. It is complex valued; it has both a real and an imaginary part. And it is, usually, a rather complicated function.

If we could describe our N-particle system classically, that is, according to classical Newtonian physics, it would be given by the set of positions and velocities for all the constituent particles:

$$[\mathbf{r}_1(t), \mathbf{r}_2(t), \ldots, \mathbf{v}_1(t), \mathbf{v}_2(t), \ldots]. \tag{1.7}$$

If each of the objects or particles that make up the system resides in a d-dimensional world, where d usually is three, this constitutes *a single point* in a real space of dimension $2Nd$.

For a quantum system, however, we need the wave function of Eq. (1.6) for a full description. This is certainly not a single point in any Nd-dimensional space; for *each and every single point* in that space, the wave function has a certain value that we need to know.

If we want to describe a function numerically, we typically need to limit the extension of the space to a finite interval and 'chop it into pieces'; we need to *discretize* it. Suppose that our numerical domain consists of s points for each of our N particles. Then our numerical, approximate wave function will consist of s^N complex numbers, while the corresponding classical state will consist of $2Nd$ real numbers. In other words, in the quantum case, the complexity grows *exponentially* with the number of particles N, while it just increases linearly with N in the classical case.

This is the infamous *curse of dimensionality*. It is quite devastating when we want to simulate quantum systems with several particles; that is really, really hard. If we want to describe a dynamical system of N quantum particles in $d = 3$ dimensions, $N = 3$ would already be too much in the usual case.[4]

For now, we limit our wave function to a single particle, such as an electron. Moreover, for the sake of simplicity, we take the world to be one-dimensional:

$$\Psi = \Psi(x; t). \tag{1.8}$$

The wave function is required to be finite, continuous and differentiable on its entire domain. This restriction is not limited to the one-dimensional one-particle case.

We already stated that the wave function contains all the information there is to be obtained about the quantum system we are interested in. However, extracting and interpreting this information may require some effort.

[4] In terms of information processing, however, this *curse* may actually be an advantage. In fact, the curse of dimensionality is what motivated the famous physicist Richard Feynman and, independently, the mathematician Yuri Manin to propose the idea of a *quantum computer* [15, 26]. More on that in Chapter 6.

1.4 Interpretation of the Wave Function

As mentioned, since a quantum particle actually is a wave, it does not really have a well-defined position or momentum – as would be the case for any kind of wave. However, a *measurement* of either position or momentum would in fact produce a definite answer. But wait, if we first measure position x and then momentum p and get definite answers, wouldn't that violate the uncertainty principle, Ineq. (1.5)? No, because when we perform a measurement, we alter the wave; the wave function prior to measurement is not the same as the one we have afterwards. If we measure the position of a quantum particle and then its momentum, a second position measurement would probably provide a result different from the first one. The momentum measurement renders the first position measurement obsolete. The uncertainty principle refers to *simultaneous* measurements; while we could have precise information about either the position or the momentum of a quantum particle, there is a lower limit to how precisely we may know both simultaneously.

For now, instead of elaborating on this most peculiar feature of the wave function, we will simply continue to examine how physical quantities are calculated from this mysterious, complex function. This really isn't fair; the issue, which typically is referred to as 'the collapse of the wave function' of 'the measurement problem', certainly deserves more attention. The fact that the wave function is affected by the action of a measurement really isn't trivial. It has been the issue of much debate – from both the philosophical and the technical points of view. And it still is. Several papers and books have been written on the subject. To some extent, we will return to this issue in the following chapters.

Instead of definite positions and momenta, the wave function only provides *probabilities* for each possible outcome of a measurement. Suppose we set out to measure the position of the particle at time t. In that case, the quantity

$$\frac{\mathrm{d}P}{\mathrm{d}x} = |\Psi(x;t)|^2 \tag{1.9}$$

provides the probability density for measuring the particle's position to be x. Or, in other words, the probability of finding the particle between positions a and b is

$$P(x \in [a,b]) = \int_a^b |\Psi(x;t)|^2 \, \mathrm{d}x. \tag{1.10}$$

Put in statistical terms: the outcome of a position measurement on a quantum particle is a stochastic variable, and $|\Psi(x;t)|^2$ is its *probability density function*. This interpretation, which was stated in a footnote in his publication, won the German physicist and mathematician Max Born the 1954 Nobel Prize in Physics.

1.4.1 Exercise: Normalization

For a quantum system consisting of a single particle in one dimension, we must insist that the following holds at all times for the wave function:

$$\int_{-\infty}^{\infty} |\Psi(x;t)|^2 \, \mathrm{d}x = 1. \tag{1.11}$$

Why?

This requirement poses some restrictions on $\Psi(x)$. Specifically, why must we insist that the wave function vanish in the limits $x \to \pm\infty$?

If we prepare lots of identical particles in the very same state, that is to say, all of the particles have identical wave functions, and then measure the position of each and every one of them, we will *not* get the same result for each measurement, rather we will have a *distribution* of results.

With sufficiently many particles and measurements, this distribution is simply Eq. (1.9) – with the proper overall scaling/normalization. This distribution has both a *mean value*, or *expectation value*, and a *standard deviation*, or *width*. With Eq. (1.9) being the probability density function for the outcome of a position measurement, the position expectation value is simply

$$\langle x \rangle = \int_{-\infty}^{\infty} x |\Psi(x)|^2 \, \mathrm{d}x. \tag{1.12}$$

This could also be written as

$$\langle x \rangle = \int_{-\infty}^{\infty} [\Psi(x)]^* \; x \; \Psi(x) \, \mathrm{d}x, \tag{1.13}$$

where the asterisk indicates complex conjugation and the time dependence of the wave function Ψ is taken to be implicit. Equation (1.13) is, of course, just a slightly more cumbersome way of writing the same thing as in Eq. (1.12); hopefully the motivation for formulating it in this way will be apparent shortly.

The *width*, or *standard deviation*, associated with our position distribution is, according to standard statistics, determined as

$$\sigma_x = \sqrt{\langle x^2 \rangle - \langle x \rangle^2}, \quad \text{where} \quad \langle x^2 \rangle = \int_{-\infty}^{\infty} x^2 |\Psi(x)|^2 \, \mathrm{d}x. \tag{1.14}$$

It is this width, σ_x, that enters into the Heisenberg uncertainty relation, Ineq. (1.5).

It is hard to avoid asking questions like 'But where actually *is* the particle prior to measurement?' and 'How come the description in terms of a wave function is unable to reveal the actual position of the particles?' And, surely, since we got different results, the various systems couldn't really have been prepared completely identically, could they?

These are, of course, bona fide and well-justified questions. As mentioned, these and similar questions have been subject to much debate since the birth of quantum physics. The most famous debates had Albert Einstein and Niels Bohr themselves as opponents. Many physicists resort to answering the above questions in the spirit of Bohr by saying something like 'It does not really make sense to talk about a particle's position without actually measuring it' and 'There is no more information obtainable beyond that contained in the wave function.'

In any case, experience does suggest that this stochastic behaviour from nature's side actually is a truly fundamental one. Although we perform identical measurements on identically prepared particles, the outcome really is arbitrary, it cannot be predicted with certainty.

Again, let's place these issues at the back of our minds and continue with our more technical approach to Ψ – for now.

1.4.2 Exercise: Position Expectation Values

Suppose that our wave function is given by one of the simple, elementary functions below, labelled from (a) to (d):

$$\Psi_a(x) = \frac{1}{(1+(x-3)^2)^{3/2}}, \tag{1.15a}$$

$$\Psi_b(x) = \frac{e^{-4ix}}{(1+(x-3)^2)^{3/2}}, \tag{1.15b}$$

$$\Psi_c(x) = e^{-x^2}, \tag{1.15c}$$

$$\Psi_d(x) = (x+\mathrm{i})\, e^{-(x-3\mathrm{i}-2)^2/10}. \tag{1.15d}$$

In each case, we can use the function to determine several physical quantities. But first we must make sure that Eq. (1.11) is fulfilled.

For each of the wave function examples, do the following:

1. Normalize it, that is, impose a factor $\Psi \to c\Psi$, such that the integral of $|\Psi(x)|^2$ over all x actually is 1.
2. Plot the absolute square of the wave function. Also, plot real and imaginary contributions separately when Ψ is complex.
3. From the plot, estimate the most likely outcome of a position measurement.
4. Determine the position expectation value.

Although some of the integrals involved may be done analytically, we strongly suggest that you write a few lines of code and calculate them numerically. This could, for instance, be done using the trapezoidal rule:

$$\int_a^b f(x)\,\mathrm{d}x = h\left(\frac{1}{2}f(x_0) + f(x_1) + \cdots + f(x_{n-1}) + \frac{1}{2}f(x_n)\right) + \mathcal{O}(h^2), \tag{1.16a}$$

$$\text{where} \quad h = \frac{b-a}{n} \quad \text{and} \quad x_k = a + kh. \tag{1.16b}$$

In doing so, you may very well use some ready-made function in your programming language or numerical platform of preference. But do make sure that you use a numerical grid which is sufficiently fine – that your h value in Eq. (1.16b) is small enough or, equivalently, that your n value is large enough. Also, instead of actually integrating from $-\infty$ to ∞, you must settle for an interval of finite extension. When doing so, make sure that your numerical domain extends widely enough; the interval $[a, b]$ must actually contain the wave function to a sufficiently high degree. In case some of your expectation values turn out complex, this may indicate that this requirement is not met.

In any case, redo your calculations several times with an increasing number of sample points n and an increasingly wide interval $[a, b]$ to ensure that your result does not rely notably on these numerical parameters.

The observations from this exercise may lead you question the relevance of any complex factor of the form e^{ikx} in the wave function. You, will, however, come to learn that it does in fact matter.

The next exercise is a direct application of Eq. (1.10).

1.4.3 Exercise: Is the Particle in This Interval?

For each of the four wave functions in Exercise 1.4.2, the *normalized* ones, what is the probability that a position measurement would determine that the particle is localized between $x = 1$ and $x = 2$?

1.5 The Momentum Operator

In quantum physics, every physical quantity that can be measured has its own corresponding linear *operator*. In this context, an operator is some specific way of changing or transforming the wave function. The operator corresponding to position we have already met – it simply consists in multiplying the wave function by x itself.

Of course, position is not the only physical quantity in town, we may very well be interested in the momentum of the particle as well. The operator for the *momentum* p, which is the product of mass and velocity, $p = mv$, is given by spatial differentiation:

$$\hat{p}\,\Psi(x) = -i\hbar\Psi'(x), \tag{1.17}$$

where $\hbar$, the so-called *reduced Planck constant*, is the Planck constant divided by 2π:

$$\hbar = \frac{h}{2\pi} \approx 1.055 \cdot 10^{-34}\ \text{J s}. \tag{1.18}$$

We use the hat symbol, which you see above the 'p' on the left hand side in Eq. (1.17), to indicate that we are dealing with an *operator*, to distinguish it from scalar quantities – numbers – and functions. We have chosen not to dress up the position operator with any hat since it simply is position itself, $\hat{x} = x$. With Leibniz's notation it makes sense to write the $\hat{p}$-operator without reference to any explicit wave function:

$$\hat{p} = -i\hbar\frac{d}{dx}. \tag{1.19}$$

Analogously to the position expectation value, Eq. (1.13), the *momentum* expectation value is

$$\langle p \rangle = \int_{-\infty}^{\infty} [\Psi(x)]^* \ \hat{p}\, \Psi(x)\, \mathrm{d}x \qquad (1.20)$$
$$= \int_{-\infty}^{\infty} [\Psi(x)]^* \left(-\mathrm{i}\hbar \frac{\mathrm{d}}{\mathrm{d}x}\right) \Psi(x)\, \mathrm{d}x = -\mathrm{i}\hbar \int_{-\infty}^{\infty} [\Psi(x)]^* \ \Psi'(x)\, \mathrm{d}x.$$

Perhaps Eq. (1.13) makes more sense now; it is identical to Eq. (1.20) – with p replaced by x, and the $\hat{p}$-operator replaced by the $\hat{x}$-operator.

Suppose we measure the momentum of a large number of particles with identical wave functions. The distribution of the results would provide the mean value of Eq. (1.20) – analogously to how the mean value of position measurements is provided by Eq. (1.12) or (1.13).

The operator for a physical quantity that depends on position and momentum also depends on the position and momentum operators in the same manner. Kinetic energy, for instance, may be written as $1/2\, mv^2 = p^2/2m$. Correspondingly, its operator is

$$\hat{T} = \frac{\hat{p}^2}{2m} = \frac{1}{2m}\left(-\mathrm{i}\hbar\frac{\mathrm{d}}{\mathrm{d}x}\right)^2 = -\frac{\hbar^2}{2m}\frac{\mathrm{d}^2}{\mathrm{d}x^2}. \qquad (1.21)$$

In the next exercise, and most of the ones following it, we will set

$$\hbar = 1. \qquad (1.22)$$

This may seem a bit overly pragmatic – if not to say in clear violation of Eq. (1.18). We know that the Planck constant is supposed to be a really small number – not 1. But it is in fact admissible. It simply corresponds to choosing a convenient set of units. And it facilitates our numerical calculations. We must be aware, though, that the numbers that come out certainly will not correspond to numbers given in 'ordinary' units such as metres or seconds.

1.5.1 Exercise: Momentum Expectation Values

For the four simple wave functions in Exercise 1.4.2, calculate the momentum expectation values. According to Eq. (1.20), this involves differentiation. You could, of course, do this by hand using paper and pencil in this case. However, it may be more convenient to use the midpoint rule:

$$f'(x) = \frac{-f(a-h) + f(a+h)}{2h} + \mathcal{O}(h^2). \qquad (1.23)$$

This way you could perform the differentiation of various wave functions without really changing anything in your code. Do remember to use *normalized* wave functions.

You may notice that for the purely real wave functions, the momentum expectation values all seem to be zero. Why is that? Can you understand that by analytical means?

Perhaps this time you are able to see a clear distinction in expectation values for the wave functions in Eqs. (1.15a) and (1.15b) – contrary to what was the case in Exercise 1.4.2?

1.6 Some Simplifying Notation

After a while you may get tired of writing up integrals. In the following we will write expressions such as Eqs. (1.12) and (1.20) somewhat more compactly. For two functions Ψ and Φ we define the notation

$$\langle \Phi, \Psi \rangle = \int_{-\infty}^{\infty} [\Phi(x)]^* \, \Psi(x) \, \mathrm{d}x. \tag{1.24}$$

With this, the expectation value of some physical quantity A for a quantum system with wave function Ψ may be written

$$\langle A \rangle = \langle \Psi, \hat{A}\Psi \rangle, \tag{1.25}$$

where $\hat{A}$ is the operator corresponding to the physical variable A, be it position, momentum, energy or something else. Analogously to Eq. (1.14), the *width* or *standard deviation* of the physical variable A is

$$\sigma_A \equiv \sqrt{\langle A^2 \rangle - \langle A \rangle^2}. \tag{1.26}$$

1.6.1 Exercise: Standard Deviations

For each of the four wave functions in Exercise 1.4.2, their *normalized* version that is, what is the standard deviation of the position, Eq. (1.14)?

And what is the momentum standard deviation?

In answering the latter, you will need to calculate expectation values for p^2, for which this finite difference formula may come in handy:

$$f''(x) = \frac{f(x-h) - 2f(x) + f(x+h)}{h^2} + \mathcal{O}(h^2). \tag{1.27}$$

Check that the uncertainty principle, Ineq. (1.5), isn't violated for any of the wave functions in question. Does any of them fulfil it with equality?

Here, the notation introduced in Eq. (1.24) simply serves as a lazy person's way of writing integrals. However, it goes deeper – deeper into linear algebra. The thing is, the definition in Eq. (1.24) may be considered an *inner product*. A general inner product fulfils

$$\langle \beta, \alpha \rangle = \langle \alpha, \beta \rangle^* \quad \text{(symmetry)}, \tag{1.28a}$$

$$\langle \alpha, c\beta \rangle = c\langle \alpha, \beta \rangle \quad \text{(linearity)}, \tag{1.28b}$$

$$\langle \alpha, \beta + \gamma \rangle = \langle \alpha, \beta \rangle + \langle \alpha, \gamma \rangle \quad \text{(linearity)}, \tag{1.28c}$$

$$\langle \alpha, \alpha \rangle > 0 \quad \text{(positivity)}, \tag{1.28d}$$

where α, β and γ are general vectors and c is a complex number. In other words, we consider functions as a kind of generalized vector in this context. We restrict these to functions that are continuous and have continuous derivatives. We also insist that the integral of their absolute value squared is finite.

The last equation above applies to all vectors in the space except the zero vector, which any vector space is required to have. The inner product between the zero vector and itself is zero.

1.6.2 Exercise: The Inner Product

Prove that the inner product between wave functions defined in Eq. (1.24) in fact fulfils the general requirements for an inner product, Eqs. (1.28).

As you may have seen, Eqs. (1.28b) and (1.28c) are satisfied by the linearity of the definite integral. As for the zero vector, we must pick the zero function, which is zero for any value of x.

1.6.3 Exercise: Hermicity

In quantum physics we insist that all operators corresponding to physical variables are *Hermitian*. This means that any operator for a physical quantity A should fulfil

$$\langle \Phi, \hat{A}\Psi \rangle = \langle \hat{A}\Phi, \Psi \rangle, \tag{1.29}$$

for all admissible wave functions Ψ.

(a) Use integration by parts to show that the operator $\hat{p}$ in Eq. (1.19) is in fact Hermitian. To this end, do remember the lesson learned at the end of Exercise 1.4.1.
(b) Show that Hermicity ensures real expectation values.
 Why is this feature crucial for quantum physics to make sense?
(c) If we impose a simple phase factor on our wave function, $\Psi \to e^{i\phi}\Psi$, where ϕ is real, this does not affect any expectation value calculated from Ψ. Why is that?

Actually, although Eq. (1.24) is a rather standard mathematical notation for inner products, it is customary to write it slightly differently in quantum physics:

$$\langle \Phi, \Psi \rangle \to \langle \Phi | \Psi \rangle \quad \text{and} \quad \langle \Phi, \hat{A}\Psi \rangle \to \langle \Phi | \hat{A} | \Psi \rangle. \tag{1.30}$$

This notation is referred to as the *Dirac notation*. It has a couple of advantages. One advantage is that it allows us to think of $|\Psi\rangle$ as a general vector; $\langle\Phi|$ is a dual vector, which in combination with $|\Psi\rangle$ forms an inner product. A $|\Psi\rangle$-type vector is referred to as a *ket* and a $\langle\Phi|$-type vector is called a *bra*, so that together they form a '*bra-ket*'. Often one may think of $|\Psi\rangle$ as a column vector and an operator $\hat{A}$ as a square matrix. In the next chapter we will see that this literally is so when we approximate the wave function on a numerical grid. In other situations it is actually true without approximation. In both cases, the bra $\langle\Phi|$ will be a row vector; it is the *Hermitian adjoint*, indicated by $\dagger$, that is, the transpose and complex conjugate, of the corresponding ket $|\Phi\rangle$,

$$\langle \Phi | = |\Phi\rangle^\dagger. \tag{1.31}$$

Later, we will see that introducing these formal tools and inner products does have advantages beyond the ability to write integrals in a more compact fashion.

1.6.4 Exercise: The Hydrogen Atom and Atomic Units

In order to get a feel for which kind of quantities we typically are dealing with for quantum systems, we calculate expectation values for an electron attached to a proton – a hydrogen atom, that is – when its energy is minimal. In this case, we may use the wave function

$$\Psi(r) = \frac{2}{\sqrt{a_0^3}} r e^{-r/a_0} \tag{1.32}$$

to calculate expectation values in the same way as for quantum particles in one dimension, as in Eqs. (1.12) and (1.20) – except that our variable now is r, the distance between the electron and the proton in the atom. Correspondingly, the lower integration limit should be zero rather than minus infinity.

When dealing with atoms, it is convenient to introduce *atomic units*, which may be defined by choosing the electron's mass m_e as our mass unit, the elementary charge e as the charge unit and the so-called *Bohr radius* a_0 as our length unit – in addition to setting $\hbar = 1$ as in Eq. (1.22). The elementary charge e is the magnitude of the charge of both the proton and the electron, the latter being negative.

In general units, the Bohr radius reads

$$a_0 = \frac{4\pi\epsilon_0\hbar^2}{m_e e^2}, \tag{1.33}$$

where the constant ϵ_0 is the so-called *permittivity of free space*. Since all other factors in the above expression are 1 in atomic units, this means that $4\pi\epsilon_0$ also equals one atomic unit.

Atomic units are usually abbreviated as 'a.u.' – irrespective of which quantity we are dealing with, it is not customary to specify whether we are talking about atomic length units or atomic time units, for instance.[5]

(a) Calculate the expectation values $\langle r \rangle$ and $\langle p \rangle$ using SI units – units following the *International System of Units*. Here, as in the one-dimensional cases above, you can take the operator for r to be r itself, and the 'p-operator', corresponding to the momentum component in the radial direction, to be $-i\hbar\, d/dr$. You will have to look up various constants.

(b) Repeat the calculations from (a) using atomic units.

The energy operator is the sum of the operator for kinetic energy, Eq. (1.21), and the potential energy. The potential energy corresponding to the attraction from the proton is

$$V(r) = -\frac{e^2}{4\pi\epsilon_0}\frac{1}{r}, \tag{1.34}$$

[5] That is not to say that this is a good practice.

so the full energy operator in this specific case may be written as[6]

$$\hat{H} = -\frac{\hbar^2}{2m_e}\frac{\mathrm{d}^2}{\mathrm{d}r^2} - \frac{e^2}{4\pi\epsilon_0}\frac{1}{r}. \tag{1.35}$$

The reason why we call it '$\hat{H}$' rather than '$\hat{E}$' is that the energy operator is named after the Irish mathematician William Rowan Hamilton.

(c) Calculate the expectation value for the energy of the system,

$$\langle \varepsilon \rangle = \langle \Psi | \hat{H} | \Psi \rangle. \tag{1.36}$$

Do so using both using SI units and atomic units.
Note: All these expectation values may be calculated analytically. However, in case you choose to do it numerically, be warned that the Coulomb potential, Eq. (1.34), diverges when r approaches zero. This could cause problems numerically – unless you let r start at a small, positive value instead of zero – or rewrite your integral a little bit before estimating it.

Hopefully, you agree that it's worthwhile to introduce a set of convenient units to replace the standard SI units. Dealing with metres and seconds just isn't that convenient when addressing atoms and other objects pertaining to the micro world. However, when you need to present your numerical results in a different context, you may have to convert your results to another set of units. If so, care must be taken to make sure that you convert your results correctly.

[6] Here we rather pragmatically jump from one dimension to a fully three-dimensional example. This is hardly ever as straightforward as this example would indicate. The reason why we can simplify it here is that the wave function depends on the distance from the origin only; it is independent of direction.

2 The Schrödinger Equation

In this chapter we will, finally, get acquainted with the famous Schrödinger equation. You may find its father, the Austrian Erwin Schrödinger, in the third row close to the middle in Fig. 1.3. The time-dependent form of his equation is a partial differential equation that tells us how the wave function evolves in time. In a compact manner it may be written

$$i\hbar \frac{\partial}{\partial t}\Psi = \hat{H}\Psi, \tag{2.1}$$

where the *Hamiltonian* $\hat{H}$ is the energy operator. In Eq. (1.35) we learned what this operator looks like for the case of a hydrogen atom at rest.[1] As mentioned, it inherited its name from the Irish mathematician William Rowan Hamilton, who made significant contributions to the theoretical foundation of classical mechanics in the nineteenth century.

For a single particle not subject to any external interactions, such as electromagnetic fields, the Hamiltonian is simply the sum of kinetic and potential energy:

$$\hat{H} = \hat{T} + \hat{V} = \frac{\hat{\mathbf{p}}^2}{2m} + V(\mathbf{r}), \tag{2.2}$$

where the potential V is a function of position. Here, momentum and position, $\mathbf{p}$ and $\mathbf{r}$, are written as vectors; that is, we are not necessarily restricted to one spatial dimension. In this case, the position operator is still just multiplication with the position itself, $\hat{\mathbf{r}} = \mathbf{r}$, while the momentum operator is proportional to a gradient operator:

$$\hat{\mathbf{p}} = -i\hbar\nabla. \tag{2.3}$$

Here '$\hat{\mathbf{p}}^2$' is taken to mean the scalar product with itself; the kinetic energy term becomes proportional to the *Laplacian*:

$$\frac{\hat{\mathbf{p}}^2}{2m} = \frac{1}{2m}\hat{\mathbf{p}} \cdot \hat{\mathbf{p}} = \frac{1}{2m}(-i\hbar\nabla) \cdot (-i\hbar\nabla) = -\frac{\hbar^2}{2m}\nabla^2. \tag{2.4}$$

In the general case, the Hamiltonian of a quantum system may be rather involved. For instance, if there are N particles that interact with each other and with an external electromagnetic field, the Hamiltonian would read

$$\hat{H} = \sum_i^N \left[\frac{(\mathbf{p} - q_i \mathbf{A}(\mathbf{r}_i, t))^2}{2m_i} + V(\mathbf{r}_i)\right] + \sum_{i>j}^N W(\mathbf{r}_i, \mathbf{r}_j). \tag{2.5}$$

[1] Truth be told, Eq. (1.35) is a bit of an oversimplification.

Here, q_i is the electric charge and m_i is the mass of particle number i; $\mathbf{A}$ is the so-called *vector potential*, which provides the electric and magnetic fields via

$$\mathbf{E} = -\partial/\partial t\ \mathbf{A} \quad \text{and} \quad \mathbf{B} = \nabla \times \mathbf{A}, \tag{2.6}$$

respectively.[2] If the system under study were an atom or a molecule with N electrons, the two-particle interaction W would be the Coulomb repulsion between electrons:

$$W(\mathbf{r}_i, \mathbf{r}_j) = \frac{e^2}{4\pi\epsilon_0} \frac{1}{|\mathbf{r}_i - \mathbf{r}_j|}, \tag{2.7}$$

and the potential $V(\mathbf{r}_i)$ would be the Coulomb attraction from the nuclei, Eq. (1.34), whose positions are frequently assumed to be fixed.[3] As mentioned, the constants e and ϵ_0 are the elementary charge and the permittivity of free space, respectively.

It is fair to say that the Hamiltonian of a composite quantum system can be fairly complicated – even with several simplifying assumptions. However, for the remainder of this chapter we will exclusively be occupied with one single particle in one dimension – without interactions with any external electromagnetic field. Our Hamiltonian will then be

$$\hat{H} = -\frac{\hbar^2}{2m} \frac{\mathrm{d}^2}{\mathrm{d}x^2} + V(x). \tag{2.8}$$

We will play around with solving Eq. (2.1) for various situations. But before we can do that, we need to make a good numerical approximation to the Hamiltonian, including the kinetic energy term above.

2.1 Numerical Discretization

We start by choosing a certain domain for our wave function Ψ to 'live' in. Assuming that $\Psi(x;t)$ vanishes for $x < a$ and for $x > b$ at all times, we may disregard space beyond these points. Moreover, as we did in Exercise 1.4.2, in Eq. (1.16b) we *discretize* the resulting interval so that our wave function may be represented by a set of points. Effectively, our approximate wave function becomes a column vector in $\mathbb{C}^{n+1}$:

$$\Psi(x;t) \to \boldsymbol{\Psi}(t) = \begin{pmatrix} \Psi(x_0;t) \\ \Psi(x_1;t) \\ \vdots \\ \Psi(x_n;t) \end{pmatrix}, \tag{2.9}$$

[2] Actually, even Eq. (2.5) is a simplification; the vector potential $\mathbf{A}$ is really a much more involved *operator* which takes into account how photons appear and vanish. For strong fields it does, however, make sense to disregard this.

[3] As nuclei are also quantum particles, they really should be amongst the N particles contained in the Hamiltonian. However, due to the fact that nuclei are much heavier than electrons, we may often assume the nuclear positions to be fixed relative to the electrons of the atom.

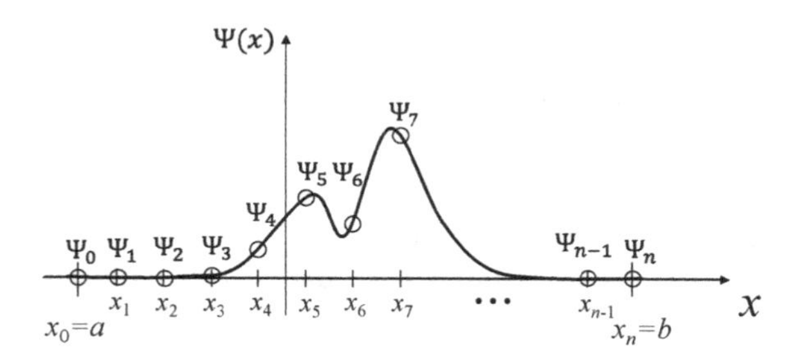

Figure 2.1 The discretization scheme of Eq. (2.9). If the interval $[a, b]$ is large enough to contain the wave function, and the grid points are chosen densely enough, the array of Ψ-values, $\Psi(x_0)$, $\Psi(x_1)$, $\Psi(x_2)$ and so on, may be used to interpolate the true wave function with reasonable accuracy. For the record, the grid displayed here is not nearly dense enough.

where $x_i = a + ih$ and $h = (b - a)/n$ as before. Please do not confuse the numerical parameter h with the Planck constant. We have tried to illustrate this scheme in Fig. 2.1.

The approximation of Eq. (2.9) is, of course, a tremendous reduction in complexity. The wave function can in principle be *any* complex function that is finite, differentiable and normalizable on $\mathbb{R}$, while in Eq. (2.9) it is reduced to a simple vector with $n + 1$ entries.

We repeat that, when imposing discretization like this, it is absolutely crucial that we check for convergence; our results should certainly not depend significantly on our choices for the numerical parameters a, b and n. We *must* increase the span of our interval $[a, b]$ and reduce our step size h, increasing n, until our numerical predictions are no longer affected by such modifications. Then, and only then, may we reasonably assume that the discretized representation $\boldsymbol{\Psi}(t)$ is a decent approximation to the true wave function $\Psi(x;t)$.

In the following we will, for convenience, skip writing out the time dependence of the wave function explicitly.

2.1.1 Exercise: Normalizing the Discretized Wave Function

When the wave function is represented as a vector with elements $\Psi(x_0), \Psi(x_1), \ldots$, we insist that

$$h \sum_{i=0}^{n} |\Psi(x_i)|^2 = h\, \boldsymbol{\Psi}^\dagger \boldsymbol{\Psi} = 1, \tag{2.10}$$

where $\dagger$ indicates the Hermitian adjoint, which, as mentioned, means transpose and complex conjugation.

Why do we insist on this?

Hint: Apply Riemann integration or the trapezoidal rule, Eq. (1.16a), to Eq. (1.11).

Hopefully you found that this was a revisit of Exercise 1.4.1 – formulated in terms of a numerical integral.

2.2 Kinetic Energy Numerically

Now, with a proper numerical discretization of Ψ we may apply numerical differentiation methods to it. The most straightforward approach would be that of *finite difference schemes*. We may, for instance, apply the three-point rule we saw in Exercise 1.6.1 or the symmetric five-point rule for the double derivative:

$$f''(x) = \frac{f(x-h) - 2f(x) + f(x+h)}{h^2} + \mathcal{O}(h^2), \text{and} \tag{2.11a}$$

$$f''(x) = \frac{-f(x-2h) + 16f(x-h) - 30f(x) + 16f(x+h) - f(x+2h)}{12h^2} + \mathcal{O}(h^4), \tag{2.11b}$$

respectively.

2.2.1 Exercise: The Kinetic Energy Operator as a Matrix

With Ψ given in vector form as in Eq. (2.9), we may write the action of $\hat{T}$ on Ψ as a matrix multiplication with the $\boldsymbol{\Psi}$ vector. For the above choices, Eqs. (2.11), what will the corresponding matrix approximations be? Assume that $\Psi(x) = 0$ for $x \notin [a, b]$ and that it falls off smoothly towards these edges.

There are, of course, several other methods to estimate differentiation of various orders numerically. A particularly convenient one is provided by the Fourier transform, which is defined as

$$\Phi(k) = \mathcal{F}\{\Psi(x)\}(k) = \frac{1}{\sqrt{2\pi}} \int_{-\infty}^{\infty} e^{-ikx}\, \Psi(x)\, \mathrm{d}x. \tag{2.12}$$

This shifts our position-dependent wave function into another function that depends on the wave number k instead. We may transform this function back into the original x-dependent function by the inverse Fourier transform, $\mathcal{F}^{-1}$:

$$\Psi(x) = \mathcal{F}^{-1}\{\Phi(k)\}(x) = \frac{1}{\sqrt{2\pi}} \int_{-\infty}^{\infty} e^{+ikx}\, \Phi(k)\, \mathrm{d}k. \tag{2.13}$$

Within the 'k-space', differentiations are trivial:

$$\begin{aligned} \frac{\mathrm{d}^n}{\mathrm{d}x^n}\Psi(x) &= \frac{\mathrm{d}^n}{\mathrm{d}x^n} \frac{1}{\sqrt{2\pi}} \int_{-\infty}^{\infty} e^{ikx}\, \Phi(k)\, \mathrm{d}k = \frac{1}{\sqrt{2\pi}} \int_{-\infty}^{\infty} \left(\frac{\mathrm{d}^n}{\mathrm{d}x^n} e^{ikx}\right) \Phi(k)\, \mathrm{d}k \\ &= \frac{1}{\sqrt{2\pi}} \int_{-\infty}^{\infty} e^{ikx}\, (\mathrm{i}k)^n\, \Phi(k)\, \mathrm{d}k = \mathcal{F}^{-1}\{(\mathrm{i}k)^n \Phi(k)\}(x). \end{aligned} \tag{2.14}$$

This means that differentiation to any order may be performed by first Fourier transforming Ψ into 'k-space', multiplying this transformed wave function by ik to the proper power and then transforming it back into the x-representation. The action of the kinetic energy operator, for instance, may be calculated as

$$\hat{T}\Psi = -\frac{\hbar^2}{2m}\mathcal{F}^{-1}\left\{(ik)^2\mathcal{F}\{\Psi\}\right\} = \frac{\hbar^2}{2m}\mathcal{F}^{-1}\left\{k^2\mathcal{F}\{\Psi\}\right\}. \tag{2.15}$$

The same approach still applies for a discretized wave function, Eq. (2.9), for which Eq. (2.12) is replaced by a discrete version. The integral over the continuous k-variable is replaced by a sum over discrete k-values and, in effect, our numerical wave function, which was defined for x between a and b, becomes a periodic function beyond this interval – with period $L = b - a$, the size of our domain.

Fortunately, discrete numerical Fourier transforms may be performed extremely efficiently, and standard implementations for the *fast Fourier transform*, FFT, are easily found within all numerical frameworks.

In discrete 'k-space', the x-vector, $(x_0, x_1, \ldots, x_n)$, is replaced by a k-vector. The maximum magnitude of k is inversely proportional to the spatial step size h:

$$k_{\text{max}} = \frac{\pi}{h}, \tag{2.16}$$

and the k-vector corresponding to the Fourier-transformed wave function extends from $-k_{\text{max}}$ to (almost) k_{max} in n steps of length

$$\Delta k = \frac{2k_{\text{max}}}{N}, \tag{2.17}$$

where $N = n + 1$ is the number of points. Note, however, that FFT implementations typically distort this k-vector in a somewhat non-intuitive manner; when N is even, it typically starts from zero and reaches $(N/2-1)\cdot\Delta k$ and then continues from $-N/2\cdot\Delta k$ to $-1\cdot\Delta k$. Check the documentation of your standard FFT implementation within your preferred framwork in order to work this out.

Actually, quantum physics may be formulated in 'k-space' or momentum space, as an alternative to the usual position space formulation. In that case, the $\hat{x}$ and $\hat{p}$ operators change roles, so to speak, and the momentum wave function $\Phi(k)$ gives the momentum probability distribution in the same manner as $\Psi(x)$ gives the position probability distribution. The momentum variable p and the wave number k are related by $p = \hbar k$, which means that they are the same in our convenient units. If you want to gain a deeper understanding of this – or how the discrete and continuous Fourier transforms relate to each other – we highly recommend the YouTube video in Ref. [1]. The video also indicates how this is related to the Heisenberg uncertanty relation, Ineq. (1.5).

2.2.2 Exercise: Expectation Values as Matrix Multiplications

(a) For the wave functions in Exercise 1.4.2, calculate the expectation value of the kinetic energy, Eq. (1.21), in the same manner as you did in Exercise 1.4.2. Do this using both a finite difference formula, Eqs. (2.11), and some FFT implementation in your preferred framework for numerics.

(b) Now, do the same but via matrix multiplication. Let your wave function be represented as in Eq. (2.9). With this, and proper normalization, the expectation value may now be estimated as

$$\langle T \rangle \approx h \, \boldsymbol{\Psi}^{\dagger} \, T \, \boldsymbol{\Psi}, \tag{2.18}$$

where $\boldsymbol{\Psi}$ is the complex column vector of Eq. (2.9) and T is a square matrix.

For each of the four wave functions, choose an adequate domain, $[a, b]$, and implement both the representations of Exercise 2.2.1 and the corresponding FFT matrix – with an increasing number of grid points, N. The FFT matrix may, as any linear transformation from $\mathbb{C}^N$, be determined by transforming the identity matrix – column by column or in one go.

Do these estimates reproduce your findings in (a)?

Are these numerical representations of the kinetic energy operator actually Hermitian – does $T = T^{\dagger}$ numerically?

The benefit of reformulating the above calculations in terms of matrix multiplications can be questioned in the above exercise. The process will, however, prove quite useful now that we are, finally, going to solve the Schrödinger equation.

2.3 Dynamics Without Explicit Time Dependence in the Hamiltonian

When the Hamiltonian $\hat{H}$ does not bear any explicit time dependence, the solution of the Schrödinger equation, Eq. (2.1), may be formally written as

$$\Psi(t) = \exp\left[-\mathrm{i}\hat{H}(t - t_0)/\hbar\right] \Psi(t_0), \tag{2.19}$$

where $\Psi(t_0)$ is the wave function at some initial time t_0 and the spatial dependence of Ψ is implicit this time. It may seem odd for an operator or a matrix to appear as an exponent. However, it does make sense if we define it as a series expansion. For any number x, it holds that

$$e^x = \sum_n^{\infty} \frac{1}{n!} x^n = 1 + x + \frac{1}{2} x^2 + \frac{1}{3!} x^3 + \cdots . \tag{2.20}$$

In the same manner, we may define the exponential function with an operator or a matrix as exponent:

$$e^{\hat{A}} \equiv \sum_{n=0}^{\infty} \frac{1}{n!} \hat{A}^n, \tag{2.21}$$

where $\hat{A}^0 = \hat{I}$ is the identity operator – the operator that does nothing.

In the following we will approximate the Hamiltonian by matrices, as we did in the preceding exercises for the kinetic energy operator. When actually calculating matrix exponentials we rarely resort to Taylor series as in Eq. (2.21); such an expansion would typically have to be truncated if implemented. However, we can exponentiate a matrix exactly if it is diagonalizable. And ours is; theory from linear algebra ensures this for any Hermitian matrix.[4] When a matrix A is Hermitian, $A = A^\dagger$, there will always exist matrices P and D such that

$$A = PDP^\dagger, \tag{2.22}$$

where D is diagonal with the eigenvalues of the matrix A, which necessarily are real, by the way, along the diagonal, $D = \mathrm{Diag}(\lambda_0, \lambda_1, \lambda_2, \ldots)$. Moreover, the matrix P is unitary; $P^\dagger = P^{-1}$.

With this we may write

$$e^{-\mathrm{i}H\Delta t/\hbar} = P\ \mathrm{Diag}(e^{-\mathrm{i}\varepsilon_0 \Delta t/\hbar}, e^{-\mathrm{i}\varepsilon_1 \Delta t/\hbar}, \ldots)P^\dagger, \tag{2.23}$$

where $\varepsilon_0, \varepsilon_1, \ldots$ are the eigenvalues of the Hamiltonian matrix H.

For obtaining these eigenvalues and the corresponding eigenvectors, the columns of P, you will certainly find adequate numerical implementations in any platform or language. However, you are also likely to find implementations that can perform the exponentiation in Eq. (2.19) directly.

2.3.1 Exercise: The Formal Solution

(a) Prove that Eq. (2.19) in fact *is* a solution of the Schrödinger equation, Eq. (2.1), with the proper initial conditions.
(b) Prove that Eq. (2.21) and Eq. (2.22) in fact lead to Eq. (2.23).

These results will be quite useful in the following, when we play around with travelling wave packets. For that we need a proper initial wave function, $\Psi(x;t_0)$. In this context, working with Gaussian wave packets is quite convenient. A general Gaussian has the form

$$f(x) \sim \exp\left[-\frac{1}{2}\left(\frac{x-\mu}{\sigma}\right)^2\right], \tag{2.24}$$

where μ is the mean position and σ is the width (see Fig. 2.2).

When the position distribution $|\Psi(x)|^2$ has this shape, μ is the mean position, Eq. (1.12), and σ is the width of the position distribution, Eq. (1.14). Gaussians are practical for several reasons. One of these is their rather simple form. And although their tails never actually vanish completely, they are, for all practical purposes, indeed confined in space[5] – and in momentum. Moreover, an initial Gaussian wave packet allowed to travel freely will remain Gaussian. Specifically, with your normalized wave function at time $t = 0$ being

$$\Psi(x, t = 0) = \sqrt{\frac{\sqrt{2}\sigma_p/\hbar}{\sqrt{\pi}(1 - 2\mathrm{i}\sigma_p^2\tau/(\hbar m))}}\ \exp\left[-\frac{\sigma_p^2(x - x_0)^2/\hbar^2}{1 - 2\mathrm{i}\sigma_p^2\tau/(\hbar m)} + \mathrm{i}p_0 x/\hbar\right], \tag{2.25}$$

[4] Perhaps this result is more familiar when it is formulated in terms of real matrices: a real symmetric matrix is always diagonalizable. The same holds for complex matrices – with symmetry replaced by Hermicity.

[5] That is to say, their magnitude quickly falls below machine accuracy as x departs from μ.

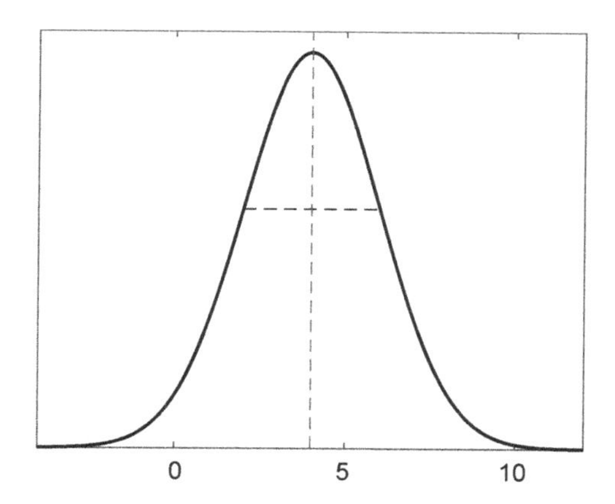

Figure 2.2 Shape of a normalized Gaussian curve, as defined in Eq. (2.24). Here the mean value $\mu = 4$, and the standard deviation $\sigma = 2$. These are illustrated by vertical and horizontal dashed lines, respectively.

the absolute value squared of our wave function at later time t is

$$|\Psi(x;t)|^2 = c^2 \exp\left[-\frac{2}{\hbar^2}\frac{\sigma_p^2(x - x_0 - p_0 t/m)^2}{1 + 4\sigma_p^4(t-\tau)^2/(m\hbar)^2}\right] \tag{2.26}$$

with the time-dependent normalization factor

$$c^2 = \sqrt{\frac{2}{\pi}}\,\frac{\sigma_p/\hbar}{\sqrt{1 + 4\sigma_p^4(t-\tau)^2/(m\hbar)^2}}.$$

Although slightly more 'messy', Eq. (2.26) is indeed of the same form as Eq. (2.24),[6] with time-dependent μ and σ. Here x_0 is the particle's mean position initially, τ is the time at which the wave packet is at its narrowest, p_0 is the mean momentum and σ_p is the width of the momentum distribution.

2.3.2 Exercise: Wave Propagation

In this exercise you are going to simulate a travelling Gaussian wave packet that is not exposed to any potential $V(x)$. With $V = 0$, the Hamiltonian is simply the kinetic energy operator, $\hat{H} = \hat{T}$ with $\hat{T}$ given in Eq. (1.21). By choosing units so that both $\hbar$ and the mass m become 1, this is quite simple: $\hat{T} = -1/2\ \mathrm{d}^2/\mathrm{d}x^2$. This system serves well for checking the accuracy of the numerical approximations since the exact solution is known analytically, Eq. (2.26).

For starters, you can choose the following set of parameters:

x_0	σ_p	p_0	τ
−20	0.2	3	5

[6] It becomes a bit 'cleaner' with $m = 1$ and $\hbar = 1$.

Also, let your domain be $[a, b] = [-L/2, L/2]$ with $L = 100$ length units. Choose an initial number of grid points $N = n + 1$, and let your initial time be $t = 0$. Because you will be using the FFT, it's an advantage to set $N = 2^k$ where k is an integer. In other words, N should be 64, 128, 256 and so on.

(a) For each of three approximations to the kinetic energy operator, the finite difference approximations of Eqs. (2.11) and the Fourier approach of Eq. (2.15), simulate the evolution of the wave function according to Eq. (2.19) and compare the numerical estimate to the exact one, Eq. (2.26). Specifically, for a set of times $t_0 = 0, t_1, t_2, \ldots$ where $t_{i+1} = t_i + \Delta t$, plot $|\Psi(x;t_i)|^2$ at each time step t_i. Preferably, do this with all four versions of the wave function – the three numerical approximations and the exact, analytical one – simultaneously as an animation. The transition from one time step to another is achieved by Eq. (2.19) and, if necessary, Eq. (2.23):

$$\Psi(t + \Delta t) = e^{-i\hat{H}\Delta t/\hbar}\, \Psi(t). \tag{2.27}$$

Choose a reasonably small Δt so that when you iterate over time as $t \to t + \Delta t$ and update your plot repeatedly, it renders a reasonably smooth animation.

(b) Play around with the numerical parameter N. For each of the three numerical estimates, how large an N do you need for your estimate to, more or less, coincide with the analytical exact solution? Which implementation seems to be the most precise one? Is the wave function in fact at its narrowest when $t = \tau$?

(c) What happens to the different approximations to the wave function when they hit the boundary at $x = \pm L/2$?

(d) Repeat this exercise playing around with various choices for the parameters in the initial Gaussian wave, σ_p, p_0, x_0 and τ.

The odd, artificial behaviour seen when our numerical approximations hit the boundary has to do with the boundary condition we happened to impose. Our FFT approximation to the kinetic energy operator requires our wave function to fulfil a *periodic boundary condition*. In the finite difference approximations, we imposed the restriction $\Psi(\pm L/2) = 0$ at all times. This boundary condition, which in effect corresponds to putting hard walls at $x = \pm L/2$, is an example of a *Dirichlet* boundary condition, named after the German mathematician Peter Gustav Lejeune Dirichlet.

The numerical tools you just developed will prove quite useful throughout the remainder of this book. Your set of numerical implementations constitutes a nice test bed for exploring several quantum phenomena.

Interference is one such phenomenon.

2.3.3 Exercise: Interference

Construct an initial wave packet of the form

$$\Psi(x, t = 0) = \frac{1}{\sqrt{2}} \left(\psi_1(x) + \psi_2(x)\right), \tag{2.28}$$

where each of the two functions ψ_1 and ψ_2 is of the form Eq. (2.25). The parameters should not be the same. Specifically, make sure to choose values for the mean momentum and initial positions such that the two Gaussians travel towards each other; the p_0 values for ψ_1 and ψ_2 must differ in sign. Also, make sure that the two initial Gaussians have negligible overlap initially. If so, the pre-factor $1/\sqrt{2}$ above ensures normalization.

Now, as in Exercise 2.3.2, simulate the evolution according to the Schrödinger equation for the system – still without any potential. In this case, just use the FFT version. What happens?

Also plot the real and imaginary parts of the wave function separately in this case, not just the absolute value squared.

Although somewhat simplified, the pattern that emerges comes about in the same manner as the interference pattern that Thomas Young saw in his famous double-slit experiment with light. Odd as it may seem, the interference phenomenon you just saw is real and measurable; the double-slit experiment has in fact been performed with particles instead of light (see Fig. 2.3). Note that, while this interference pattern does correspond to two interfering waves, it does *not* correspond to two particles interfering with each other. There is only one quantum particle, and it *interferes with itself*.

2.3.4 Exercise: Expectation Values and Uncertainties

Choose a Gaussian wave packet, Eq. (2.26), with $\tau > 0$ and use the FFT implementation from Exercise 2.3.2 to make a plot of the expectation values $\langle x \rangle$ and $\langle p \rangle$ as functions of time.

Also, calculate the uncertainty, or *standard deviation*, of the position and momentum as functions of time. You will find relevant relations in Eqs. (1.12), (1.20) and (1.26). In determining $\langle p \rangle$ and $\langle p^2 \rangle$, you will need to estimate $\mathrm{d}\Psi(x)/\mathrm{d}x$ and $\mathrm{d}^2\Psi(x)/\mathrm{d}x^2$, respectively. To this end, both Eqs. (1.23) and (2.11a) or Eq. (2.14) will do.

Finally, confirm numerically that the uncertainty in momentum remains constant and that the Heisenberg uncertainty principle, Ineq. (1.5), is not violated.

Does equality in Ineq. (1.5) apply at any time? If so, when?

A small reminder: when we set the reduced Planck constant $\hbar = 1$, this means that the Planck constant $h = 2\pi$.

Hopefully, you found that the Schrödinger equation makes sure that this wave function abides by the uncertainly principle at all times. This will also be the case for any other admissible wave function.

So far we have let our wave function evolve freely. And not much has happened to it. It would be more interesting to let our wave functions hit something. It is about time we introduced a *potential* in our Hamiltonian.

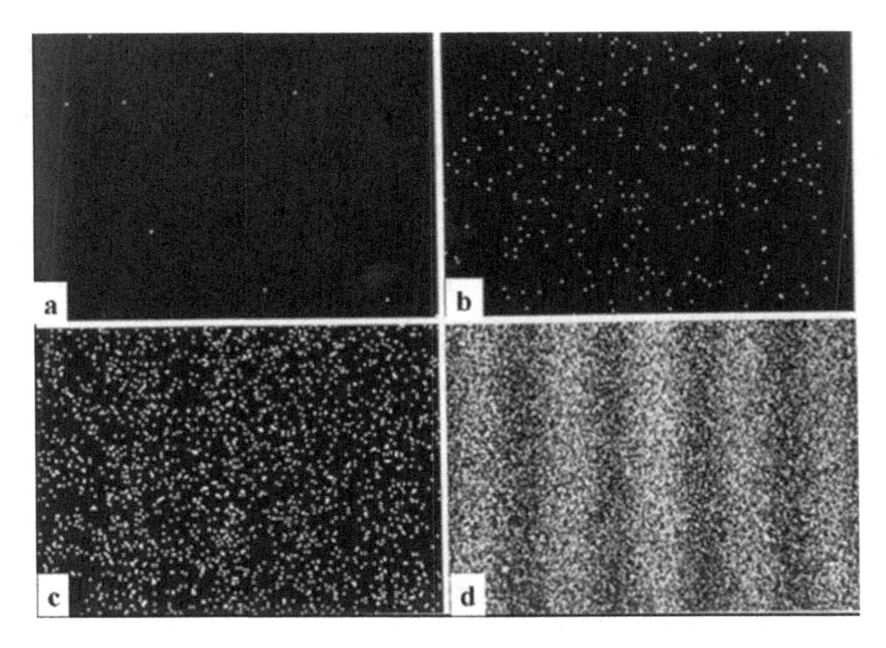

Figure 2.3 The famous double-slit experiment, which Thomas Young performed with light back in 1801 (see Fig. 1.1), can be done with *particles* as well. Electrons are sent through narrow slits, one by one. The spots appear where electrons hit an electron-sensitive screen afterwards. We see that, as more and more particles are detected, an interference pattern emerges. Specifically, the numbers of electrons in the pictures are (a) 8, (b) 270, (c) 2000 and (d) 160,000 [2]. The original experiment, published in 1989, was conducted by A. Tonomura, J. Endo, T. Matsuda and T. Kawasaki, at the Hitachi Advanced Research Laboratory, and H. Ezawa, at Gakushuin University, Japan [39]. A similar experiment was performed by the Italians P. G. Merli, G. F. Missiroli and G. Pozzi in 1976 [28, 36]. The first observation of interference between electrons, however, was made by Clinton Davisson and Lester Germer, who in the 1920s studied how electrons scattered off a nickel surface.

2.4 Scattering and Tunnelling

We now introduce a potential V in the Hamiltonian. For simplicity, we let this potential be a rectangular one, centred at the middle of the grid:

$$V(x) = \begin{cases} V_0, & |x| \leq w/2, \\ 0, & |x| > w/2, \end{cases} \tag{2.29}$$

where w is the width. The value V_0 can be positive, in which case the potential is a barrier, or negative, in which case it may confine our quantum particle. For now we address the former case.

Actually, for numerical simulations, instead of the purely rectangular potential above, we will use a smoother version:

$$V_s(x) = \frac{V_0}{e^{s(|x|-w/2)} + 1}. \tag{2.30}$$

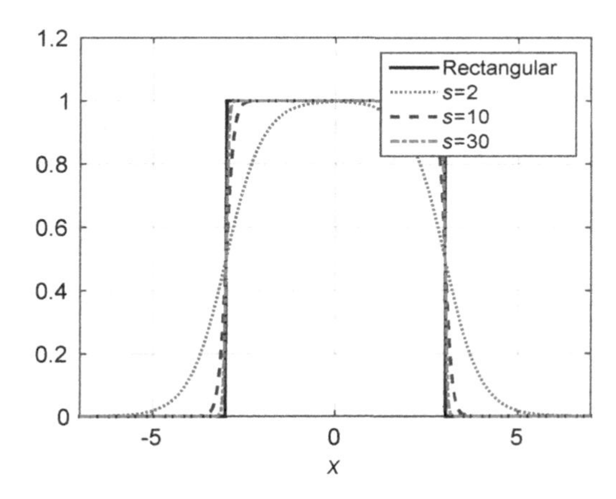

Figure 2.4 The rectangular potential in Eq. (2.29) – along with the 'smooth' version of Eq. (2.30) for three different values of s. Here $V_0 = 1$ and $w = 6$.

This one is a bit more convenient to work with numerically. The parameter s fixes the smoothness. In the limit $s \to \infty$ we reproduce the sharp corners of the potential in Eq. (2.29). The potential is plotted for various s values in Fig. 2.4 – along with the purely rectangular one.

2.4.1 Exercise: Scattering on a Barrier

In this exercise you do not get to choose the parameters yourself – at least not initially. In your implementation from Exercise 2.3.2, again with the initial wave function, or initial *state*, as in Eq. (2.26), you introduce a potential of the above shape, Eq. (2.30). This is easy, you just augment your Hamiltonian with the potential V, which becomes a diagonal matrix in our representation:

$$\text{Diag}(V(x_0), V(x_1), V(x_2), \dots) = \begin{pmatrix} V(x_0) & 0 & 0 & \\ 0 & V(x_1) & 0 & \cdots \\ 0 & 0 & V(x_2) & \\ & \vdots & & \ddots \end{pmatrix}, \tag{2.31}$$

before you perform the exponentiation, Eq. (2.27). Again, we suggest you use the FFT representation of kinetic energy.

Use the following parameters:

L	$n+1$	σ_p	p_0	x_0	τ	V_0	w	s
200	512	0.2	1	−20	0	3	2	5

As in Exercise 2.3.2, make an animation/simulation of where your wave packet hits the barrier. You may want to indicate the position of the barrier somehow in your simulation.

What happens?

After the wave packet has collided with the barrier, what is the final probability for the wave function to remain on the left side of the barrier ($x < 0$)? And what is the probability that it has passed to the right side of it ($x > 0$)? The former is called the *reflection probability*, R, and the latter is called the *transmission probability*, T:

$$R = \int_{-\infty}^{0} |\Psi(x)|^2 \, dx, \tag{2.32a}$$

$$T = \int_{0}^{\infty} |\Psi(x)|^2 \, dx = 1 - R, \tag{2.32b}$$

where these quantities should, strictly speaking, be calculated in the limit $t \to \infty$. This would, however, require a grid of infinte extension. In practice, you determine these quantities for a time T that is long enough for the collision with the barrier to be over but short enough to prevent the numerical wave function hitting the boundary at $x = \pm L/2$.

Rerun the scattering process with different choices for p_0 and V_0. How are the reflection and transmission probabilities affected by these adjustments?

Finally, replace the V_0 value with a negative one so that the barrier becomes a well instead.

Will your quantum particle, with some probability, fall into the well and get trapped? Do you still get reflection with a negative V_0 in the potential?

We suppose that the answer to the last question is affirmative. Would a classical particle – a particle that follows Newton's laws – behave this way? We could check.

2.4.2 Exercise: The Dynamics of a Classical Particle

It could be interesting to compare the wave function with the position that a classical particle would have. If we include this in the same simulation as in Exercise 2.4.1, it will serve to illustrate some of the profound differences between classical physics and quantum physics.

Simulate numerically the trajectory of a classical particle, $x(t)$, with the initial conditions given by the mean position and mean momentum of the initial Gaussian wave packet; set $x(t = 0) = x_0$ and let the initial velocity be $v(t = 0) = p_0/m$. The classical evolution is dictated by Newton's second law, Eq. (1.1), in one dimension:

$$m x''(t) = -\frac{d}{dx} V(x). \tag{2.33}$$

If we write down separate but coupled equations for the momentum and position,

$$m \frac{d}{dt} x(t) = p \quad \text{and} \quad \frac{d}{dt} p(t) = -V'(x), \tag{2.34}$$

we may formulate the evolution as a first-order ordinary differential equation:

$$\frac{\mathrm{d}}{\mathrm{d}t}\begin{pmatrix} x \\ v \end{pmatrix} = \begin{pmatrix} v \\ -V'(x)/m \end{pmatrix}, \tag{2.35}$$

with v being the particle's velocity. This allows us to resolve the dynamics as a first-order ordinary differential equation, an *ODE*, by standard schemes such as Runge–Kutta methods for instance. You may very well make use of standard ODE routines in your numerical framework of preference. The differentiation of the potential could be done, for instance, by Eq. (1.23), or it could be done using paper and pencil.

Include a numerical solution of Eq. (2.35) along with your solution of the Schrödinger equation in your implementation from Exercise 2.4.1. Indicate the position of the classical particle along with the evolution of the quantum wave.

Now, suppose a ball is rolling uphill with an initial velocity of $v_0 = 4$ m/s. The top of the hill is at height $H = 1$ m above the ball's initial position. Why can we be absolutely certain that the ball will *not* make it to the other side of the hill? Because energy conservation prohibits it. The ball with mass $m = 1$ kg would need an energy of $mgH \approx 10$ J to reach the top of the hill, while it only has $1/2\ mv_0^2 = 8$ J to spend. Thus, for sure, the classical ball will roll down again on the same side it came from; there is no way we would retrieve the ball on the other side.

Let's have a look at the same process within the quantum world.

2.4.3 Exercise: Tunnelling

Start out with the same scenario as in Exercise 2.4.1 with the same parameters – except for one thing: this time, let $V_0 = 1$.

Again, run the simulation and see what happens. And, again, calculate the reflection probability and transmission probability, Eqs. (2.32). What is the probability for this quantum physical 'ball' to make it to the other side of the hill?

Feel free to include the implementation of Newton's second law, the one from Exercise 2.4.2, along with your solution of the Schrödinger equation in your simulation.

So, despite the fact that our quantum ball has a mean energy of about $p_0^2/(2m) = 0.5$ in our units, it still has a significant probability to be found on the other side of a 'hill' that it would take an energy of $V_0 = 1$ energy units to climb. Isn't that rather odd?

Well, you could argue that 0.5 is only the *mean* energy of the particle – more or less. Since the wave that is a quantum particle typically has a *distribution* in momentum, rather than one specific momentum, it also has a distribution in energy. And this distribution may very well contain some higher energy components which *could* make it over the barrier. If we think of the particle as a wave in water, parts of it could 'splash' over, so to speak.

This is, indeed, a bona fide argument. However, in the above case, the probability of measuring an energy beyond 1 is actually just 1.6%.[7] As seen in the exercise, the probability of turning up on the other side of the barrier is considerably higher than this. So, the oddity remains.

Let's play around with this phenomenon, which is called *tunnelling*, a bit more.

2.4.4 Exercise: Tunnelling Again

Now, we replace the barrier with two narrow barriers, symmetrically placed on either side of $x = 0$. This can be achieved with this potential:

$$V(x) = V_s(x - d) + V_s(x + d), \tag{2.36}$$

where V_s refers Eq. (2.30). Here the separation d must be significantly larger than the width w of the barriers. Set $d = 10$ length units and $w = 0.5$. Also, let $V_0 = 1$ and $s = 25$.

Make your initial state a Gaussian wave packet well localized between the barriers, but not too narrow. This time, let both the initial mean position x_0 and momentum p_0 be zero.

Then run your simulation again. What happens? Can you contain the wave functions between the barriers? Perhaps the evolution of the wave function is best studied with a logarithmic y-axis in this case.

As you go along, calculate the probability of finding the particle between the barriers as a function of time:

$$P_{\text{between}} = \int_{-d}^{d} |\Psi(x;t)|^2 \, dx, \tag{2.37}$$

and plot it after the simulation.

Simulate until a significant part of the wave packet hits the boundary at $x = \pm L/2$. You may want to increase your box size L and let your simulation run a bit longer than in the case of Exercise 2.4.3. Also, make sure to use a dense enough grid to get a decent approximation of your barriers, which are rather narrow this time.

Try to see how the tunnelling probability is affected by imposing slight adjustments on the height and the width of your barriers.

We have now seen examples of situations in which a quantum physical object, such as an electron or an atom, penetrates regions in which it does not have sufficient energy to be and then comes out on the other side.

This tunnelling phenomenon has no analogue in classical physics.[8] In effect, this means that one can actually make electrons escape an atom or some piece of metal

[7] In Exercise 5.3.3 we will see how such probabilities may be determined.

[8] A silly thought: suppose sheep were quantum physical objects. In that case, just building fences wouldn't do; the sheep would eventually tunnel through them. Instead, the farmer would have to dig a sharp depression in the ground in order to confine the herd. If they did so, however, it would get even weirder – again, if sheep were quantum objects, that is.

Figure 2.5 Image of a graphite surface obtained using a scanning tunnelling microscope. The structure you can make out is made up of individual atoms; this is, in fact, a picture of atoms!

although they do not really have sufficient energy to make this happen – in the classical sense. One very useful application of this phenomenon is the *scanning tunnelling microscope*, which actually allows us to *see* individual atoms (see Fig. 2.5). This technology, which we will return to in Chapter 6, exploits the fact that the tunnelling probability is very sensitive to the width and the height of the barrier being tunnelled through. Hopefully, this fact resonates with your experience by the end of Exercise 2.4.4.

2.5 Stationary Solutions

As we have already seen, quantum physics has some strange, non-intuitive features. Yet another one comes into play if we ask ourselves the following question: are there solutions of the Schrödinger equation for which the probability density, $|\Psi(x;t)|^2$, does not change in time? Such a situation would be reminiscent of standing waves of a vibrating string. The answer is yes – if we have a confining potential.

2.5.1 Exercise: Enter the Eigenvalue Problem

Suppose that a quantum system exposed to a time-independent Hamiltonian $\hat{H}$ is such that $|\Psi(x;t)|^2$ is also time independent.

(a) Verify that this is achieved if the system's wave function is of this form:

$$\Psi(x,t) = e^{-i\varepsilon t/\hbar}\,\psi(x). \tag{2.38}$$

Why must the constant ε be real here?

(b) Show that when we insert the wave function of Eq. (2.38) into the Schrödinger equation, Eq. (2.1), we arrive at the equation

$$\hat{H}\psi(x) = \varepsilon\psi(x) \tag{2.39}$$

for the time-independent part $\psi(x)$.

(c) Explain, without assuming a separable form as in Eq. (2.38) from the outset, how the restriction that $|\Psi(x;t)|^2$ is time independent in fact *leads to* Eq. (2.38).

Be warned, this last part of the exercise is hardly trivial. It may be done in the following way. First, note that since $|\Psi(x;t)|$ is time independent, the wave function may be written in polar form as $\Psi(x;t) = \psi(x)e^{i\varphi(x;t)}$, where both functions ψ and φ are real and ψ is time independent. Then, insert Ψ in this form into Eq. (2.1) with the Hamiltonian of Eq. (2.8) and move purely x-dependent terms to one side of the equation in order to achieve a partial separation of variables. Finally, insist that φ remains real in time.

The next chapter is dedicated to Eq. (2.39). We may recognize it as an *eigenvalue* equation. The Hamiltonian $\hat{H}$ is the operator corresponding to energy, and the eigenvalue ε on the right hand side is, in fact, energy. For each *eigen-energy* ε there is a corresponding eigenfunction or *eigenstate* $\psi_\varepsilon(x)$.

Before we direct our full attention to Eq. (2.39), which will be the topic of the next chapter, we should say a few more words about eigenvalues and eigenfunctions and how they are related to *measurement*.

2.6 Eigenstates, Measurements and Commutators

The following is a fundamental postulate of quantum physics:

A precise measurement of the energy of a quantum system will necessarily produce an eigenvalue of the energy operator $\hat{H}$ as a result.

In the next chapter we will see that this imposes severe limitations on the energies a quantum system may be observed to have when the system is confined, or *bound*.

The fact that we can only measure eigenvalues extends to *all* observables in quantum physics, not just energy. The possible outcomes of measuring *any* physical quantity A will necessarily be eigenvalues a of the corresponding operator $\hat{A}$,

$$\hat{A}\varphi_a(x) = a\varphi_a(x). \tag{2.40}$$

2.6.1 Exercise: The Standard Deviation of an Eigenstate

For a quantum system in the above eigenstate, $\Psi(x) = \varphi_a(x)$, show that the expectation value of A is a and that its standard deviation is zero.

Suppose now that we have made a measurement of some generic physical quantity A and got the result a. This has altered the wave function; it is now an eigenfunction of the $\hat{A}$-operator corresponding to the eigenvalue that was measured: $\Psi(x) \to \varphi_a(x)$. That is to say, as the system is subject to some measurement, it no longer follows the evolution dictated by the Schrödinger equation, it collapses into an eigenstate – the eigenstate φ_a corresponding to the result a of the measurement. This is the *collapse of the wave function*, which we briefly discussed in Section 1.4.

It is not for us to know in advance which eigenstate it collapses into. We can only determine the *probability* of measuring the outcome a when making an A measurement at time t:

$$P_a = |\langle \varphi_a | \Psi(t) \rangle|^2 . \tag{2.41}$$

An identical measurement on another identically prepared system may very well produce a different result. Repeated A measurements in quick succession on the same system will, however, keep on reproducing the result since the wave function Ψ collapsed into the corresponding eigenstate at the first measurement.

So why do we resort to such an odd idea as an instantaneous, random collapse of the wave function? Because it agrees with how nature actually turns out to behave. Every time we have asked her whether she really behaves like this, or in less poetic terms: whenever we have performed an experiment addressing this question, she has answered *yes*.

In Fig. 2.3, for instance, we see this collapse manifested in the fact that electrons, which initially are described as waves spreading out in space, are detected as tiny bright spots on a plate – a position measurement has caused the wave function to collapse into a localized spike. A single position measurement alone does not reveal any wave behaviour. But when we perform measurements on several electrons, each one with more or less the same wave function, and aggregate the outcomes, the wave interference pattern emerges.

Again, this quantum phenomenon has no analogue in classical physics.

Well, actually, let's try to find one. Maybe we could think of a quantum system as an odd guitar – an imaginary *quantum guitar*. When you pick a string on an actual guitar, you set it vibrating (Fig. 2.6). This vibration can be thought of as a bouquet of standing waves – or *stationary solutions*. Each wave has a wavelength following the formula $\lambda_n = L/(2n)$, where n is a positive integer and L is the length of the string. The higher the n, the more nodes has the standing wave and the shorter is its wavelength λ_n.

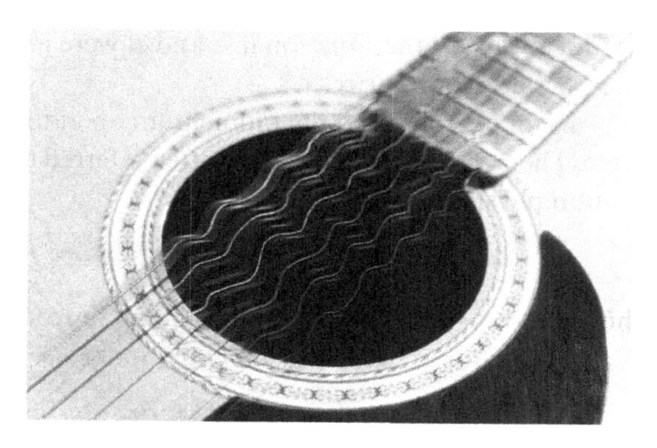

Figure 2.6 Waves on real, classical guitar strings. In case you were wondering, the picture *is* manipulated.

According to the relation $f\lambda = v$, where f is the frequency and v is the velocity of the wave,[9] there is a corresponding set of frequencies $f_n = 2nv/L$. So, when you picked the string, you didn't just produce a sound with the fundamental tone, the 'ground frequency' $f_1 = 2v/L$, you also produced a set of *harmonics* of higher frequency. The admixture of these waves, each one with its own amplitude, is what gives the guitar its characteristic sound, its *timbre*.

Now, for a *quantum guitar* the oddity enters when we wish to measure the frequency of the sound. The vibrating string on a quantum guitar could, like an ordinary guitar, initially have a distribution between various possible frequencies – or energies, see Eq. (1.4). However, a quantum measurement would produce one, and only one, of the frequencies f_n. And after measuring it, the string would vibrate as a single standing wave – with frequency f_n exclusively. This frequency is, perhaps, most likely to be the ground frequency f_1, but we will not necessarily get this result. In any case, gone is the bouquet of harmonics that gives it its particular timbre. The quantum string subject to frequency measurement would sound like a naked, cold sinusoidal tone.

Admittedly, addressing similarities and differences between classical particles and quantum particles in terms of strings on a guitar may come across as a bit far fetched. But the analogy between standing waves on a string and stationary solutions for the atom did actually inspire both de Broglie and Schrödinger. Perhaps some of the parallels between a quantum system and this odd, imaginary guitar will become more clear in the next chapter.

Before we get there, we will introduce the *commutator*, which is a rather important notion in quantum physics for several reasons. We will briefly address a couple of those reasons.

2.6.2 Exercise: Commuting with the Hamiltonian

The *commutator* between two operators $\hat{A}$ and $\hat{B}$ is defined as

$$[\hat{A}, \hat{B}] = \hat{A}\hat{B} - \hat{B}\hat{A}. \tag{2.42}$$

If this commutator happens to be zero, we say that the operators $\hat{A}$ and $\hat{B}$ *commute*. This would always be the situation if $\hat{A}$ and $\hat{B}$ were just numbers, but not necessarily if they are operators or matrices.

The position operator and the momentum operator, Eq. (1.19), for instance, do not commute. Their commutator, which is often referred to as the *fundamental commutator* in quantum physics, is

$$[\hat{x}, \hat{p}] = i\hbar. \tag{2.43}$$

(a) Show that Eq. (2.43) is correct. To this end, introduce a test function, say, $f(x)$, for the commutator $[x, \hat{p}]$ to act upon and make use of the product rule:

$$(f(x) \cdot g(x))' = f'(x)g(x) + f(x)g'(x). \tag{2.44}$$

[9] This velocity, in turn, depends on the thickness and tension of the string.

In the following, we will, by rather generic means, deduce a couple of fundamental results related to the commutators with the Hamiltonian. In this particular context we will exploit Dirac notation, Eq. (1.30), to simplify expressions – with a rather abstract outline as a consequence.

(b) For the Hamiltonian of Eq. (2.8), show that

$$[x, \hat{H}] = \mathrm{i}\frac{\hbar}{m}\hat{p} \quad \text{and} \tag{2.45a}$$

$$[\hat{p}, \hat{H}] = -\mathrm{i}\hbar V'(x). \tag{2.45b}$$

Do make use of Eq. (2.43) and the fact that any operator commutes with itself – to any power. The latter has the consequence that the position operator commutes with the potential energy part of the Hamiltonian. In proving Eq. (2.45a), the following commutator relation may also be useful:

$$[\hat{A}, \hat{B}^2] = \hat{B}[\hat{A}, \hat{B}] + [\hat{A}, \hat{B}]\hat{B}. \tag{2.46}$$

In proving Eq. (2.45b), you may want to introduce a test function $f(x)$ – as in the part (a).

(c) If we write the Schrödinger equation in Dirac notation, that is, with the wave function expressed as a ket vector, it reads $\mathrm{i}\hbar|\partial\Psi/\partial t\rangle = \hat{H}|\Psi\rangle$, where the ket $|\partial\Psi/\partial t\rangle$ is the time derivative of the $|\Psi\rangle$-ket. If we impose the Hermitian adjoint to the Schrödinger equation, we get the formal expression for the corresponding bra vector, see Eq. (1.31):

$$-\mathrm{i}\hbar\left\langle\frac{\partial\Psi}{\partial t}\right| = \left(\mathrm{i}\hbar\frac{\partial}{\partial t}|\Psi\rangle\right)^\dagger = \left(\hat{H}\,|\Psi\rangle\right)^\dagger = \langle\Psi|\,\hat{H}, \tag{2.47}$$

where we have used the fact that $\hat{H}$ is Hermitian and that the factors of a product change order under Hermitian adjungation, $(\hat{A}\hat{B})^\dagger = \hat{B}^\dagger\hat{A}^\dagger$.

With these expressions, Eqs. (2.45) and the product rule, we can show that the time derivatives of the expectation values of position and momentum follow

$$m\frac{\mathrm{d}}{\mathrm{d}t}\langle x\rangle = \langle p\rangle \quad \text{and} \quad \frac{\mathrm{d}}{\mathrm{d}t}\langle p\rangle = -\langle V'(x)\rangle. \tag{2.48}$$

Do so.

Note that this coincides with Eq. (2.34) if we replace the classical position and momentum with the corresponding quantum physical expectation values.

(d) Analogously to what you just did, show that for a physical quantity A, with operator $\hat{A}$ and no explicit time dependence, the time evolution of its expectation value follows

$$\frac{\mathrm{d}}{\mathrm{d}t}\langle A\rangle = \frac{1}{\mathrm{i}\hbar}\langle[\hat{A}, \hat{H}]\rangle. \tag{2.49}$$

Suppose $\hat{A}$ commutes with the Hamiltonian, what consequences would this have?

(e) Assume now that all eigenvalues a of some operator $\hat{A}$, see Eq. (2.40), are such that each of them only corresponds to one eigenvector $|\varphi_a\rangle$; no two linearly independent eigenvectors correspond to the same eigenvalue.[10] Assume also that some other operator $\hat{B}$ commutes with $\hat{A}$.

In this case, the eigenvectors of $\hat{A}$ would also be eigenvectors of $\hat{B}$. Why?

So, what physical insights can we take with us from these analytical efforts? First of all, we see that the Schrödinger equation actually reproduces Newton's second law if we interpret classical position and momentum as the corresponding quantum physical expectation values. This theorem, which may be generalized to higher dimensions and time-dependent Hamiltonians, is due to Paul Ehrenfest, whom you can seen in the back row, third from the left in Fig. 1.3.

Another lesson learned – from the more general and more abstract formulation of Ehrenfest's theorem in Eq. (2.49) – is that the expectation value of any quantity that commutes with the Hamiltonian will remain unchanged. Actually, we have already seen an example of this. In Exercise 2.3.2 our Hamiltonian did not feature any potential, and, consequently, it commuted with the momentum operator. Thus, according to our above abstractions, our Gaussian wave packet should evolve without any change in the momentum expectation value – or standard deviation, for that matter. Hopefully, your findings from Exercise 2.3.4 confirm this.

Another, less obvious consequence of Eq. (2.49) is that if your initial state happens to be an eigenstate of some operator $\hat{A}$ that commutes with the Hamiltonian at all times, it will also remain an eigenstate of $\hat{A}$. In the context of stationary solutions, Eq. (2.39), this means that $\hat{A}$ and $\hat{H}$ may have common eigenstates.

[10] If the opposite were the case, we would say that the eigenvalue a is *degenerate*.

3 The Time-Independent Schrödinger Equation

At the end of the last chapter, we learned that, in order to find the possible stationary states of a quantum system, we must solve the equation

$$\hat{H}\psi = \varepsilon\,\psi. \tag{3.1}$$

As this is the eigenvalue equation of the energy operator, the *Hamiltonian*, these eigenvalues are the possible outcomes of an energy measurement.

This equation is known as the *time-independent Schrödinger equation*. It is hard to guess how much time and effort researchers have spent trying to solve it for various physical and chemical systems over the years. But it's a lot, really a lot.

Although we no longer have any time dependence in our solutions, solving Eq. (3.1) certainly may still be complicated enough. However, it is not so hard in the case of one particle in one dimension, in which it reads

$$\left[-\frac{\hbar^2}{2m}\frac{\mathrm{d}^2}{\mathrm{d}x^2} + V(x)\right]\psi(x) = \varepsilon\,\psi(x). \tag{3.2}$$

3.1 Quantization

Suppose the potential V is such that it is negative in some region and that it vanishes asymptotically, $V(x) \to 0$ when $|x| \to \infty$. In this case, a particle with positive energy, $\varepsilon > 0$, is able to escape, while a particle with negative energy, $\varepsilon < 0$, is not – it is confined by the potential. We say that it is *bound*.

3.1.1 Exercise: Bound States of a Rectangular Well

For the purely rectangular well, Eq. (2.29) with $V_0 < 0$, we may find the negative eigenenergies by almost purely analytical means. This can be done by following the recipe below. The setup is illustrated in Fig. 3.1.

1. Choose adequate numerical values for how deep, $-V_0$, and wide, w, the well should be. You will not need these values right away, however; treat these quantities algebraically for now.
2. Divide the domain into three regions: region I: $x < -w/2$, region II: $|x| \leq w/2$ and region III: $x > w/2$.

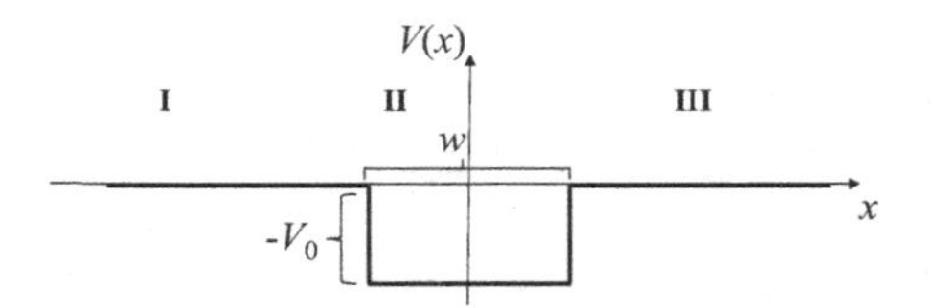

Figure 3.1 The example under study in Exercise 3.1.1.

3. Write the general solutions of the time-independent Schrödinger equation, Eq. (3.2), in each of the three regions separately. Remember that the energy ε is supposed to be negative; we are interested in *bound states*. In this context it is convenient to introduce

$$\kappa = \frac{1}{\hbar}\sqrt{-2m\varepsilon} \quad \text{in regions I and III,} \tag{3.3a}$$

$$k = \frac{1}{\hbar}\sqrt{2m(\varepsilon - V_0)} \quad \text{in region II.} \tag{3.3b}$$

Do apply our usual units where $\hbar = m = 1$ if you wish. Note that both κ and k are real; we suggest that you write your expressions for the solution in terms of real, not complex, functions.

4. You should now have a function containing six undetermined parameters. Since the wave function must be finite and normalizable, the function must vanish as $|x| \to \infty$. In order to achieve this, two of these parameters must be set to zero right away. Moreover, it can be shown that, since $V(x)$ is symmetric, the wave function has to be either even, $\psi(-x) = \psi(x)$, or odd, $\psi(-x) = -\psi(x)$.[1] We will exploit this; let's choose an even solution for now. This will reduce your number of parameters to two.

5. Impose the restrictions that both ψ and its derivative, $\psi'(x)$, are to be continuous at $x = w/2$. With a wave function that is either even or odd, these restrictions will also be fulfilled at $x = -w/2$. This will provide you with a homogeneous set of two linear equations in two unknowns. Let's write this system as a matrix equation.

   Now, here come the big questions:

   Why must we insist that the determinant of its coefficient matrix is zero?

   And how does this impose restrictions on the energy ε?

6. Make a plot of the determinant from point 5 as a function of ε and estimate which eigenenergies ε_n are admissible.

7. Repeat from point 4 with an anti-symmetric wave function.

[1] This is a manifestation of what we saw in the last question in Exercise 2.6.2; since the symmetry operator, which changes x into $-x$, commutes with our Hamiltonian, they have common eigenstates.

Perhaps this has been a bit of a laborious exercise – and perhaps it will appear even more so after having done the next one. However, it may very well turn out to be worthwhile. We just experienced that the requirement that any energy measurement on our bound quantum particle *must* render an energy eigenvalue, in turn, reduces the possible outcomes to just a few very specific ones. The notion that the time-independent Schrödinger equation in fact reduces the accessible energies for a bound particle to a discrete set is neither technically nor conceptually intuitive.

What we just saw will happen to *any* quantum system that is bound. Its energy can only assume one out of a discrete set of values. The energy is *quantized*. This is yet another quantum trait that has no classical analogue.[2]

And this is why we call it *quantum* physics.

Since the time-independent Schrödinger equation, Eq. (3.1), is, in fact, an eigenvalue equation, we could also solve it numerically by applying standard numerical tools.

3.1.2 Exercise: Solution by Direct Numerical Diagonalization

(a) Use your implementation from Chapter 2 to construct a numerical Hamiltonian for the same system as in the previous exercise. Use the potential of either Eq. (2.29) or Eq. (2.30) with rather sharp edges – a high s value, that is. Use some numerical implementation to find the eigenenergies and eigenvectors of this Hamiltonian. How do the negative eigenenergies compare with the ones you estimated in Exercise 3.1.1?

(b) Plot the wave functions, that is, the eigenvectors you found in (a).[3] Is there a pattern as you increase the energy? Do you see any parallels with the quantum guitar of Section 2.6 here?

For the bound states, you will find that the wave function has an increasing number of nodes as you increase the energy towards zero – just like the standing waves on a guitar with increasing frequency. This is illustrated in Fig. 3.2.

Now, what about situations where the system isn't bound? Well, the energy is still supposed to be an eigenvalue of the Hamiltonian, but this time it may assume *any* value – as long as it corresponds being unbound. In general, the set of eigenvalues, the *spectrum*, of a Hamiltonian will consist of both discrete values – corresponding to bound states, and continuous values – corresponding to unbound states. Of course, in special cases, it could be exclusively discrete, as would be the case when $V(x) \to \infty$ as $|x| \to \infty$, or the spectrum may be a pure continuum, as is the case when there is no potential at all – as in Exercise 2.3.2, and when the potential is too narrow or shallow to support bound states.

[2] Allow us to propose yet another silly illustration of the absurdity of quantum physics: The notion that all but a countable set of energies are prohibited for our bound particle is as reasonable as an official approaching Usain Bolt after the 100 m final at the 2009 athletics championship in Berlin and telling him, 'We regret to inform you that 9.58 s is not an admissible time, the time *must* be an integer.'

[3] In order to make sure that everything is right, you could compare your analytical solutions from Exercise 3.1.1 with your numerical ones; for a more or less rectangular potential, they should not differ by more than an overall normalization factor.

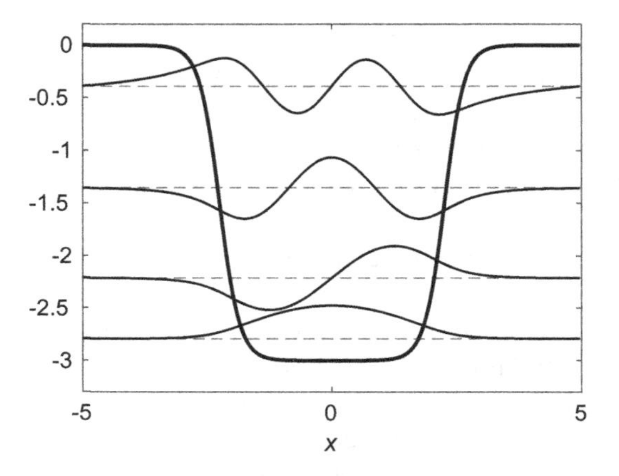

Figure 3.2 The four bound states of the smoothly rectangular potential of Eq. (2.30) with depth given by $V_0 = -3$, width $w = 4.5$ and smoothness $s = 5$. The thick curve is the confining potential, and the dashed horizontal lines are the eigen-energies of the bound states. Note that the number of nodes increases with the energy, and that, correspondingly, the wave functions, which are all real, are alternating even and odd functions.

For the case of the confined electron in a hydrogen atom, the allowed energies follow a particularly simple pattern:

$$E_n = -\frac{B}{n^2}, \tag{3.4}$$

where the constant $B = 2.179 \times 10^{-18}$ J, and n, the so-called *principal quantum number*, *must* be a positive integer. This is the famous *Bohr formula*. With $n = 1$, the energy you arrive at should coincide with what you found in Exercise 1.6.4. The set of allowed energies in Eq. (3.4) is predicted by the time-independent Schrödinger equation, in three dimensions, with the Coulomb potential:

$$V(r) = -\frac{e^2}{4\pi\varepsilon_0}\frac{1}{r}. \tag{3.5}$$

We will not derive the Bohr formula here, but you can find its derivation in virtually any other textbook on quantum physics.

The validity of Bohr's formula was well known from experiment at the time Schrödinger published his famous equation. It bears Niels Bohr's name because he had earlier arrived at this expression based on a model for the hydrogen atom which, as mentioned, is long since abandoned. He was right about two things, though.

1. The energy of an atom is indeed quantized.
2. The atomic system may 'jump' from a state of lower energy to one of higher energy by *absorbing* a photon with energy corresponding to the energy difference, see Eq. (1.4). Correspondingly, it may 'jump' to a lower lying state by *emitting* the excess energy in the form of a photon.

3.1.3 Exercise: The Bohr Formula

In Exercise 1.32 we cheated and played around with a three-dimensional hydrogen wave function as if it were one-dimensional. If we, once again, allow ourselves to take the same shortcut, we should be able to arrive at the same bound-state energies as in Eq. (3.4) rather easily. To this end, construct a numerical approximation of the Hamiltonian (1.35). Let r extend from 0 to some upper limit L. It is absolutely crucial that your implementation ensures that

$$\psi(0) = 0. \tag{3.6}$$

Achieving this with your FFT version of the kinetic energy is not so straightforward. Instead you could use the finite difference formula of Eq. (2.11a). Since $V(r)$ diverges as $r \to 0$, $r = 0$ should be omitted in your grid – as in Exercise 1.6.4. This does not prevent you from imposing the boundary condition of Eq. (3.6) in your finite difference representation of the kinetic energy. With your first grid point being $r_1 = h$, with h being the separation between neighbouring grid points, you impose the restriction $\psi(0) = 0$ when you set

$$\frac{\mathrm{d}^2\psi(r_1)}{\mathrm{d}r^2} \approx \frac{0 - 2\psi(r_1) + \psi(r_2)}{h^2} \tag{3.7}$$

in your code.

This time, contrary to Exercise 1.6.4, allow yourself to use atomic units exclusively, in which case the constant B in Eq. (3.4) is 1/2.

To what extent do the eigenenergies you obtain by numerical means coincide with the Bohr formula?

Do you see any relation between the number n, the energy *quantum number*, and the shape of the wave function? Does this compare to our silly quantum guitar?

For which n does your numerical eigenenergies start deviating from the exact ones? Is this onset manifested in the wave function somehow?[4]

How can you improve on the coincidence between the numerical and the exact eigenenergies for high n values?

So, the energy of the hydrogen atom, or, strictly speaking, the energy of the electron stuck to a proton in a hydrogen atom, is quantized. This holds for any atom – also atoms with several electrons. However, the discretized energy levels are not nearly as simple as Eq. (3.4) in any other case.

It's worth repeating that the measurement of *any* physical observable, not just the energy, will produce an eigenvalue of the corresponding operator as a result. Correspondingly, energy is not the only quantity that may be quantized. Take for instance the angular momentum operator in three dimensions:

$$\hat{\boldsymbol{\ell}} = \mathbf{r} \times \hat{\mathbf{p}} = -\mathrm{i}\hbar \mathbf{r} \times \nabla. \tag{3.8}$$

[4] With the question being posed in this way: of course it is.

If we express it – or, more conveniently, its square – in spherical coordinates, we will find that its eigenfunctions[5] only make sense physically if the eigenvalues of $\hat{\ell}^2$ are $\ell(\ell+1)\hbar^2$, where $\ell = 0, 1, 2, \ldots$.

Yet another example of a quantized quantity is *charge*. As was demonstrated in the oil drop experiment by Robert A. Millikan and Harvey Fletcher in 1909, charge is only seen in integer values of the elementary charge, $q = ne$, in nature.

We have seen that the electronic energy levels of a hydrogen atom follow a particularly simple pattern, Eq. (3.4). The same may be said about the energy of the *harmonic oscillator*.

3.1.4 Exercise: The Harmonic Oscillator

The potential

$$V(x) = \frac{1}{2}kx^2, \tag{3.9}$$

where k is some real constant, is called the *harmonic oscillator potential*. It appears in classical physics when you want to describe a mass attached to a spring, in which case k is the stiffness of the spring. In such a situation, the object in question will, according to Newton's second law, oscillate back and forth with the angular frequency

$$\omega = \sqrt{k/m}. \tag{3.10}$$

In quantum physics, this potential is also given much attention – for several reasons. Many of these, in turn, are related to the fact that the eigenenergies of the Hamiltonian with this potential are such that the differences between neighbouring energies are always the same, they are *equidistant*:

$$E_n = (n + 1/2)\,\hbar\omega, \tag{3.11}$$

where n is a non-negative integer. Note that it only has discrete, positive eigen-energies, no continuum – and that the lowest eigenenergy is not zero.

In the following we will set $m = 1$, as usual, and also $k = 1$.

Implement the Hamiltonian with the potential in Eq. (3.9) numerically and find its eigenvalues and eigenstates. Choose a domain extending from $x = -L/2$ to $x = L/2$ and a number of grid points n large enough to reproduce the exact energies, Eq. (3.11), for the first 100 energies or so. Your FFT version of the kinetic energy should be perfectly adequate here.

The harmonic oscillator potential is in fact a very popular one in textbooks as it lends itself particularly well for determining the eigenvalues of the Hamiltonian by analytical means. Its relevance is, however, certainly not limited to that.

In the following we will learn that certain energies can be prohibited also in the absence of confinement.

[5] These eigenfunctions are called *spherical harmonics*, they appear frequently in quantum physics.

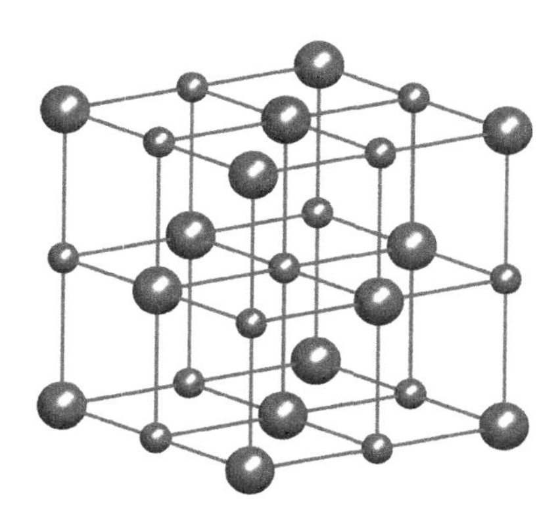

Figure 3.3 A salt crystal. The potential experienced by an electron set up by the ions in fixed positions is periodic – in several directions.

3.2 A Glimpse at Periodic Potentials

The case of periodic potentials,

$$V(\mathbf{r}) = V(\mathbf{r} + \mathbf{a}_i), \tag{3.12}$$

merits particular attention. Here $\mathbf{a}_i$ is one out of, possibly, several vectors in which $V(\mathbf{r})$ is periodic. A particle being moved by the vector $\mathbf{a}_i$ – or an integer times $\mathbf{a}_i$ – will experience exactly the same potential as before this displacement. This would, for instance, be the situation that an electron within a ionic lattice would experience (see Fig. 3.3). Here the points of the lattice would be atoms, ions or molecules that are fixed in space.

Indeed, electrons exposed to a periodic potential is a rather common situation. Solid matter is, in fact, built up of such regular patterns. We can learn very general – and relevant – lessons from studying the case of periodic potentials.

Here we will consider a single particle in one dimension exposed to a periodic Gaussian potential.

3.2.1 Exercise: Eigenenergies for Periodic Potentials – Bands

We will investigate the eigenvalues for a particle in this periodic potential:

$$V(x) = -V_0 e^{-x^2/(2\sigma^2)} \quad \text{for} \quad x \in [-a/2, a/2] \quad \text{with} \quad V(x+a) = V(x). \tag{3.13}$$

It is illustrated in Fig. 3.4.

(a) With a periodic potential, the physics must reflect the periodicity. Specifically, the solutions of the time-independent Schrödinger equation must fulfil

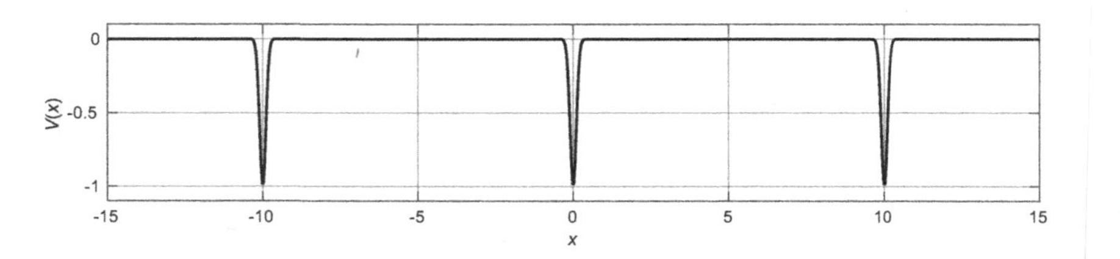

Figure 3.4 Plot of the periodic potential of Eq. (3.13) for $V_0 = -1, \sigma = 0.1$ and $a = 10$. Although we have only plotted three periods here, it extends indefinitely.

$|\psi_n(x)|^2 = |\psi_n(x+a)|^2$; the position probability distribution must be periodic in the same manner. Explain why this leads to this requirement:

$$\psi(x) = e^{i\varphi(x)}u(x), \tag{3.14}$$

where $u(x)$ is periodic with period a and $\varphi(x)$ is a real function.

Actually, it does not really make sense that the wave function would change its dependence on x if we shift x by a period, $x \to x + a$. The only way to avoid this is to insist that $\varphi(x)$ is a linear function in x:

$$\varphi(x) = \kappa x + b. \tag{3.15}$$

In this case, shifting x by a would only amount to a global phase factor, which is immaterial, as we learned in Exercise 1.6.3 (c). For the same reason, the factor b is irrelevant, and we arrive at

$$\psi(x) = e^{i\kappa x}u(x), \quad u(x+a) = u(x). \tag{3.16}$$

Note that this is not something we have proved properly; we have only tried to justify why it should make sense.[6]

(b) With Eq. (3.16), explain how the time-independent Schrödinger equation, Eq. (3.2), may be rewritten in terms of $u(x;\kappa)$ as

$$\left[-\frac{\hbar^2}{2m}\left(\frac{d}{dx} + i\kappa\right)^2 + V(x)\right]u(x;\kappa) = \varepsilon(\kappa)\, u(x;\kappa), \tag{3.17}$$

where ε and u depend parametrically on κ.

(c) Now, solve Eq. (3.17) numerically by constructing a numerical Hamiltonian and diagonalizing it – as in Exercise 3.1.2. This differs, however, from what we did in that exercise in two ways:

[6] A formal proof could follow these lines: (1) Introduce the displacement or *translation* operator $\hat{D}(d)$, which shifts the argument of a function: $\hat{D}(d)\psi(x) = \psi(x+d)$. (2) Since $\hat{D}(a)$ does not affect the Hamiltonian, the eigenfunctions of $\hat{H}$ are also eigenfunction of $\hat{D}(a)$, $\hat{D}(a)\psi(x) = \lambda\psi(x)$; see the discussion following Exercise 2.6.2. (3) With $\hat{D}(a)$ acting on $\psi(x)$ written as in Eq. (3.14), we must also have $\hat{D}(a)\psi = \exp[i\varphi(x+a)]u(x)$. (4) For these two expressions for $\hat{D}(a)\psi$ to be consistent, we must insist that $\lambda \exp[i\varphi(x)] = \exp[i\varphi(x+a)]$ – for all x. This can only hold if Eq. (3.15) holds.

(1) this time quantization comes about by the fact that $u(x)$ is periodic and

(2) the eigenenergies depend on the κ-parameter.

This means that Eq. (3.17) must be solved several times – for various values of κ.

For a few of the lowest eigenvalues $\varepsilon_n(\kappa)$, find them numerically and plot them. You can use the parameters given in Fig. 3.4.

Due to the periodicity in the wave function, the energies $\varepsilon_n(\kappa)$-s are also periodic – with period $2\pi/a$. So it is sufficient to calculate $\varepsilon_n(\kappa)$ for $\kappa \in [-\pi/a, \pi/a]$. The requirement that the eigenstates $u_n(x;\kappa)$ are periodic in x invites you to set the box size L equal to a and impose periodic boundary conditions.

Hint: $(d/dx + i\kappa)^2 u(x) = \mathcal{F}^{-1}\left\{(ik + i\kappa)^2 \mathcal{F}\{u\}(k)\right\}(x)$.

(d) Each of the eigen-energies $\varepsilon_n(\kappa)$ occupies a specific interval of the energy axis if you include all possible values of κ. For each of the energies you plotted in (c), determine, approximately, the corresponding energy interval.

The result that the eigenstates of the Hamiltonian of a periodic potential have the form of Eq. (3.16), which also extends to higher dimensions, is known as *Bloch's theorem*. The Swiss, later also US citizen, Felix Bloch arrived at the result in 1929. This theorem is far from the only important result in physics that carries his name.

In Fig. 3.5 we display the first five eigenvalues of Eq. (3.17) with the parameters listed in Fig. 3.4 as functions of κ. Such relations between an energy ε and a wave number κ are called *dispersion relations*. We see that each eigenvalue, through its κ-dependence, defines a range on the energy axis. In this context, such a range is called a *band*. In a one-dimensional system like this one, such bands do not overlap – although they could come close.

Between the bands there are *gaps*. For instance, in our example, you will never find a particle with energy between -0.01 and 0.05 energy units, nor are any energies between 0.15 and 0.2, 0.40 and 0.44, or 0.74 and 0.79 available to our particle (see Fig. 3.5). In other words, the mere fact that the potential is periodic causes certain energy intervals to be inaccessible.

Several of these aspects carry over to the more realistic situation in which we have several particles in three dimensions. In three dimensions, the structure is more complex, and bands may, contrary to the one-dimensional case, overlap. But the fact that there are bands and gaps prevails in any system featuring periodicity. In combination with the *Pauli principle*, which we will introduce in the next chapter, this has very practical consequences. Whether or not a solid may easily transmit heat or carry electrical current is very much dependent on whether the last occupied band is completely filled or not. Being excited to a higher energy would allow an electron to carry current, to be *conducting*. This does not come about so easily if such an excitation requires jumping a large energy gap to reach the next band, but it is quite feasible if it can be done within the band.

This is a *very* superficial outline of how conductance properties of solid matter emerge.

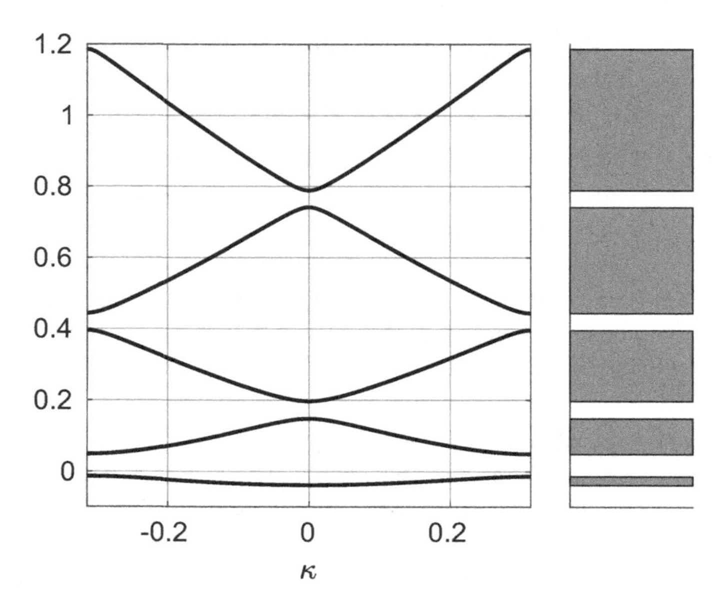

Figure 3.5 The first five eigenvalues of Eq. (3.17) for the periodic potential displayed in Fig. 3.4 as functions of κ. The right panel illustrates the corresponding energy bands and gaps.

3.3 The Spectral Theorem

There is an important result from linear algebra that is used *very* frequently in quantum physics. Actually, we have already used it – in Eq. (2.22). If we, for notational simplicity, assume that the set of eigenvalues of the Hamiltonian are entirely discrete/countable, we may write the eigenvalue equation for the energy, that is, the time-independent Schrödinger equation, as

$$\hat{H}\psi_n = \varepsilon_n \psi_n, \tag{3.18}$$

with n being an integer. Since the Hamiltonian is Hermitian, $\hat{H} = \hat{H}^\dagger$, its eigenvectors/functions form an orthogonal basis for the space in which the wave function lives. If we also require the eigenvectors/functions to be normalized under the inner product (1.24), we have *orthonormality*:

$$\langle \psi_m | \psi_n \rangle = \delta_{m,n}, \tag{3.19}$$

where the *Kronecker delta function* $\delta_{m,n}$ is 1 when $m = n$ and zero otherwise.

The fact that the eigenvectors form a basis means that *any* admissible wave function may be written as a linear combination of these:

$$\Psi = \sum_n a_n \psi_n, \tag{3.20}$$

where the coefficients a_n are uniquely determined.

Although this chapter is dedicated to the time-independent Schrödinger equation, we will briefly address time dependence also in this context. We have seen that eigenstates of a time-independent Hamiltonian are stationary in the sense that the wave function does not change in time – except for a trivial phase factor, see Eq. (2.38). However, suppose our initial state is not such an eigenstate. In that case, these phase factors may become rather important.

3.3.1 Exercise: Trivial Time Evolution

Suppose that the normalized set of eigenfunctions of the Hamiltonian, $\psi_0, \psi_1, \ldots$, is known and that your initial state is $\Psi(t=0) = \sum_n a_n \psi_n$.

(a) Why must we insist that

$$\sum_n |a_n|^2 = 1? \tag{3.21}$$

(b) Explain why

$$a_n = \langle \psi_n | \Psi(t=0) \rangle. \tag{3.22}$$

(c) Explain why

$$\Psi(t) = \sum_n a_n e^{-i\varepsilon_n t/\hbar} \psi_n. \tag{3.23}$$

Let's apply this knowledge to determine the evolution of a specific quantum system.

3.3.2 Exercise: Glauber States

We return to the harmonic oscillator of Exercise 3.1.4. This time we will investigate how the wave function for such a system moves around when we start out with a state that is not an eigenstates. One particular way of constructing such linear combinations is rather interesting.

(a) Construct some linear combination of the eigenstates:

$$\Psi(x;t=0) = \sum_{n=0}^{n_{\max}} a_n \psi_n(x), \tag{3.24}$$

that is, just pick some values a_n for a bunch of n values. Then, simulate the corresponding evolution of the modulus squared of your wave function as dictated by Eq. (3.23).

(b) Linear combinations of this particular form are called *coherent states* or *Glauber states*:[7]

$$\Psi_{\text{Glauber}} = e^{-|\alpha|^2/2} \sum_{n=0}^{\infty} \frac{\alpha^n}{\sqrt{n!}} \psi_n. \tag{3.25}$$

Here, α is some complex number.

Now repeat exercise (a) with this particular linear combination for several choices of α.

Three things are crucial for the numerical version to work:

(1) In a finite space, such as our numerical representation, we cannot sum up to infinity. Moreover, we may get into trouble numerically if we use *all* our numerical eigenstates. We suggest that you use the comparison between your numerical eigenvalues with the exact ones, Eq. (3.11), to determine a reasonable truncation in your numerical version of Eq. (3.25); by that we mean that we set all a_n to zero for all n beyond some maximum number. Since the coefficients fall off quite rapidly due to the factorial in the denominator, this should be quite admissible.

(2) The eigenfunction you get from the numerical calculation is hardly normalized according to Eq. (2.10). You probably need to impose the factor $1/\sqrt{h}$ to ensure proper normalization.

(3) As long as the eigenstates are constructed by a numerical implementation of diagonalization, we do not control the overall sign or phase of the eigenvectors. However, in order to construct the coherent state, we need to control the phase convention of the ψ_n. This may be achieved by insisting that $\psi_n(x)$ is real and positive for small, positive values of x. After your numerical diagonalization, you could run through your eigenstates, check and flip the sign where necessary.

When you run your simulation with the particular linear combination of Eq. (3.25), do you notice anything special? Try playing around with other values of α, including complex and purely imaginary ones. Use a finer grid and/or a larger domain if you have to.

(c) It is worthwhile to simulate the solution for a classical particle along with the quantum mechanical one, as in Exercise 2.4.2. In doing so, you let the real and imaginary part of α, respectively, fix the initial conditions for position and velocity:

$$x(t=0) = \sqrt{\frac{2\hbar}{m\omega}}\,\mathrm{Re}(\alpha) \quad \text{and} \quad v(t=0) = \sqrt{\frac{2\hbar\omega}{m}}\,\mathrm{Im}(\alpha). \tag{3.26}$$

The potential of Eq. (3.9) is such that it supports bound, discrete states only. As mentioned, the spectrum would feature both a discrete and a continuous part in the general case. Still, the notation in terms of a countable set, Eq. (3.20), is always adequate when

[7] Roy J. Glauber was a theoretical physicist from the USA. He made significant contributions to our understanding of light. For this he was awarded the 2005 Nobel Prize in Physics. His coherent states, Eq. (3.25), certainly had something to do with that.

we approximate the wave function in a discretized manner,[8] Eq. (2.9). In this case the wave function is just an array in $\mathbb{C}^{n+1}$ and all operators are square, Hermitian matrices. Such a description does not support any actual continuous set of eigenstates, a *continuum*. We have to settle for a so-called *pseudo-continuum*. The proper description in terms of an infinite-dimensional space within a basis that is not countable is not compatible with purely numerical representations.

We will not enter into these complications here, save to say that often we may 'get away' with considering the real physics as the limiting case where n and $b - a$ in Eq. (2.9) both become large. As mentioned, convergence in these numerical parameters is absolutely necessary for calculations to be trustable.

3.4 Finding the Ground State

Although in principle a quantum system could be found to have any of its admissible energies, or, as we say, be in any admissible *state*, systems tend to prefer the state with the lowest energy, ε_0. The *ground state*. In a somewhat superficial manner we may say that this is due to nature's inherent affinity for minimizing energy. A slightly less superficial account: eigenstates are indeed stationary and a system starting out in an eigenstate will remain in that state – unless it is exposed to some disturbance. However, avoiding disturbances, such as the interaction with other atoms or some background electromagnetic radiation, is virtually impossible. Due to such perturbations, the eigenstates are not really eigenstates in the absolute sense. While they would be eigenstates of the Hamiltonian of the isolated quantum system, they are not eigenstates of the *full* Hamiltonian, the one that would take the surroundings into account. If a system starts out in a state of higher energy, an *excited state*, such small interactions would usually cause the system to relax into states with lower energy, for instance by emission of a photon.

It is thus fair to say that the ground state of a quantum system is more important than the other stationary states, the *excited states*. Calculating ground states is crucial for understanding the properties of the elements and how they form molecules. Suppose, for instance, that the lowest energy of a two-atom system is lower when the atoms are separated just a little bit than if they are far apart; the system has lower energy if the atoms are together than if they are not. This means that these atoms would form a stable molecule.

It wasn't until people started solving the time-independent Schrödinger equation and calculating ground states that the periodic table started to make sense. There was no doubt that Dimitri Mendeleev's arrangement of the elements was a meaningful one, but it took some time to learn *why* (see Fig. 3.6).

The spectral theorem provides two very powerful methods for estimating the wave function and the energy of the ground state.

[8] Provided that the numerical grid is large and dense enough, that is.

ОПЫТЪ СИСТЕМЫ ЭЛЕМЕНТОВЪ.

ОСНОВАННОЙ НА ИХЪ АТОМНОМЪ ВѢСѢ И ХИМИЧЕСКОМЪ СХОДСТВѢ.

			Ti = 50	Zr = 90	? = 180.
			V = 51	Nb = 94	Ta = 182.
			Cr = 52	Mo = 96	W = 186.
			Mn = 55	Rh = 104,4	Pt = 197,1.
			Fe = 56	Rn = 104,4	Ir = 198.
			Ni = Co = 59	Pl = 106,6	Os = 199.
H = 1			Cu = 63,4	Ag = 108	Hg = 200.
	Be = 9,4	Mg = 24	Zn = 65,2	Cd = 112	
	B = 11	Al = 27,1	? = 68	Ur = 116	Au = 197?
	C = 12	Si = 28	? = 70	Sn = 118	
	N = 14	P = 31	As = 75	Sb = 122	Bi = 210?
	O = 16	S = 32	Se = 79,4	Te = 128?	
	F = 19	Cl = 35,5	Br = 80	I = 127	
Li = 7	Na = 23	K = 39	Rb = 85,4	Cs = 133	Tl = 204.
		Ca = 40	Sr = 87,6	Ba = 137	Pb = 207.
		? = 45	Ce = 92		
		?Er = 56	La = 94		
		?Yt = 60	Di = 95		
		?In = 75,6	Th = 118?		

Д. Менделѣевъ

Figure 3.6 The original periodic table that Dimitri Mendeleev published in 1869. It has evolved a bit since then.

3.4.1 Exercise: The Variational Principle and Imaginary Time

(a) Suppose we start out with some arbitrary admissible, normalized wave function. Why can we, by exploiting Eq. (3.20), be absolutely sure that the expectation value of its energy cannot be lower than the ground state energy?

How can this be exploited in order to establish an upper bound for the ground state energy?

(b) Again, suppose that we start out with some arbitrary wave function. Now we replace the time t in the time-dependent Schrödinger equation with the imaginary time $-\mathrm{i}t$. This corresponds to the replacement $\Delta t \to -\mathrm{i}\Delta t$ in Eq. (2.27). This, in turn, leads to a 'time evolution' in which the norm of the wave function is *not* conserved. To compensate for that, we also renormalize our wave function at each time step when we resolve the 'time evolution' in small steps of length Δt. In other words: at each step we multiply our wave function by a factor so that its norm becomes 1 again.

Why does this numerical scheme, in exponential time, turn virtually *any* initial state into the ground state?

How can the change in norm at each time step be used in order to estimate the ground state energy?

Hint: Assume that your wave function, after some 'time' t, already is fairly close to the ground state so that $\hat{H}\Psi \approx \varepsilon_0\Psi$.

Note that when we make the replacement $t \to -it$, this effectively turns the (time-dependent) Schrödinger equation into a sort of *diffusion equation*. This equation, however, does not describe any actual dynamics; it is merely a numerical trick used to determine ground state energies.

Both methods alluded to here are frequently used within quantum physics and quantum chemistry – maybe the approach of part (a) more than the imaginary time-approach in part (b). The principle that the ground state is a lower bound to the expectation value of *any* admissible function,

$$\langle \Psi | \hat{H} | \Psi \rangle \geq \varepsilon_0, \tag{3.27}$$

is called the *variational principle*. We can exploit it by constructing some initial guess for the ground state wave function, a *trial function*, with tunable parameters. The closer this initial guess is to including the actual ground state wave function, the better. And the more parameters you have to 'tweak', the more flexibility you have to achieve a good estimate of the ground state energy. The lower the energy you get, the closer you get to the ground state energy. On the other hand, the more parameters you have, the more cumbersome is the optimization you need to run. Therefore, it is crucial to choose your trial functions as cleverly as possible.

The Norwegian Egil Hylleraas (Fig. 3.7), was particularly clever in this regard. He was one of the pioneers when it comes to variational calculations. In fact, he was head

Figure 3.7 Norwegian physicist Egil Hylleraas. At a time when the time-independent Schrödinger equation was known to explain the structure of the hydrogen atom, Hylleraas succeeded in proving that it is also able to explain the much more complicated helium atom. This was a very significant contribution at the time. His methods are still quite relevant within quantum chemistry.

hunted, so to speak, by Max Born to show that the time-independent Schrödinger equation indeed produced the right ground state energy for the helium atom, not just the hydrogen atom. At a time when quantum physics was not yet established as *the* proper framework for atomic and molecular physics, it was crucial to settle this matter. Hylleraas succeeded in doing so. The fact that he did so before researchers had access to modern-day computer facilities only serves to render his achievements even more impressive.

Now, let's do some variational calculations ourselves – *with* a computer:

3.4.2 Exercise: One-Variable Minimization

Let your potential be provided by Eq. (2.30) with $V_0 = -3$, $s = 5$ and $w = 8$. Also, take your trial wave function to be a normalized Gaussian centred at the origin:

$$\psi_{\text{trial}} = \frac{1}{\sqrt[4]{2\pi\sigma^2}} \exp\left[-\frac{x^2}{4\sigma^2}\right]. \tag{3.28}$$

Now, we calculate the energy expectation value,

$$E(\sigma) = \langle \psi_{\text{trial}} | \hat{H} | \psi_{\text{trial}} \rangle, \tag{3.29}$$

for various values of σ. Do this numerically using the same implementation of the Hamiltonian as before and plot the expectation value of the energy as a function of σ. Try to read off its minimal value – and the corresponding σ.

In this case, actually obtaining the 'correct' ground state energy is easy by means of direct diagonalization, which we did in Exercise 3.1.2. Do this again with the parameters at hand and compare the actual ground state energy to the upper bound you just found. Also, compare the numerical ground state from your trial function, Eq. (3.28), with the σ that minimizes $E(\sigma)$.

Hopefully, you found that both the energy and the wave function that minimized the energy were reasonably close to the correct ones. Hopefully, you will also find that you can do better with the method of imaginary time.

3.4.3 Exercise: Imaginary Time Propagation

For the same potential as in Exercise 3.4.2, implement the method in Exercise 3.4.1(b), in which an arbitrary initial state is propagated in imaginary time and the wave function is normalized at each time step:

$$\psi_{n+1} = e^{-\hat{H}\Delta t/\hbar}\psi_n, \quad \psi_{n+1} \to \psi_{n+1} \Big/ \sqrt{\langle \psi_{n+1} | \psi_{n+1} \rangle}. \tag{3.30}$$

Choose a reasonably small 'time step', $\Delta t = 0.05$ time units, perhaps, and plot the renormalized, updated wave function at the first few steps. Also, write the estimated energy to screen for each iteration.

How does this implementation compare to the variational principle implementation above? Were you able to reproduce the very same ground state and ground state energy as you found by direct diagonalization? Did that happen with the variational calculation in Exercise 3.4.2?

The fact that the method of imaginary time estimates the ground state better does not mean that variational calculations are irrelevant and inferior. Often, variational calculations allow us to obtain quite good results when addressing problems of high complexity. And the flexibility of the method may be improved by adding more parameters to our trial wave functions.

3.4.4 Exercise: Two-Variable Minimization, Gradient Descent

This time, let your initial 'guess' be a Gaussian with *two* parameters, μ and σ:

$$\psi_{\text{Gauss}} = \frac{1}{\sqrt[4]{2\pi\sigma^2}} \exp\left[-\frac{(x-\mu)^2}{4\sigma^2}\right]. \tag{3.31}$$

Now estimate the ground state energy, or, strictly speaking, determine an upper bound for the ground state energy, for this potential:

$$V(x) = V_s(x-2) + x^2/50, \tag{3.32}$$

where you fix the parameters $V_0 = -5$, $w = 6$ and $s = 4$ for V_s in Eq. (2.30).

As your energy estimate now depends on two parameters, $E(\mu, \sigma)$, minimization is not as straightforward as in Exercise 3.4.2. In order to obtain values for both μ and σ which minimize the expectation value of the energy, apply the method of *gradient descent* or *steepest descent*, as it is also called.

Gradient descent is a powerful method for minimizing a general, differentiable multivariable function,[9] $F(\mathbf{x})$. Here $\mathbf{x}$ is a vector contaning all the variables that F depends on. You start out by choosing a starting point $\mathbf{x}_0$. Next, you move one step in the direction in which the function descends the most steeply – in the direction opposite to the gradient:

$$\mathbf{x}_{n+1} = \mathbf{x}_n - \gamma \nabla F(\mathbf{x}_n). \tag{3.33}$$

The factor γ, which is called the *learning rate* when applied within machine learning, fixes the length of the step made for each iteration – together with the magnitude of the gradient. For optimal performance, γ may be adjusted as you go along. However, with a sufficiently small γ value, this should not be crucial in our case.

The iterations of Eq. (3.33) are repeated as long as $F(\mathbf{x}_{n+1}) < F(\mathbf{x}_n)$ – or as long as the decrease is significant enough for each iteration. Finally, you should arrive very close to some local minimum.

[9] In our case, this would be $E(\mu, \sigma)$.

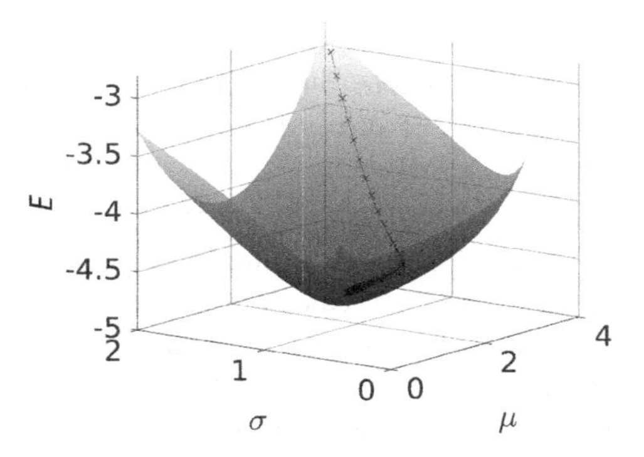

Figure 3.8 The surface illustrates how the energy expectation value of the trial wave function, Eq. (3.31), depends on the parameters μ and σ. It is a friendly landscape with only one minimum, no other pesky local minima in which we could get stuck. The dashed path towards the minimum is the result of a gradient descent calculation starting with $\mu = 4, \sigma = 2$ with the learning rate γ fixed at 0.1.

The partial derivatives that make up the gradient can be estimated numerically using the midpoint rule,[10] Eq. (1.23):

$$\frac{\partial}{\partial \mu} E(\mu,\sigma) \approx \frac{E(\mu+h,\sigma) - E(\mu-h,\sigma)}{2h},$$
$$\frac{\partial}{\partial \sigma} E(\mu,\sigma) \approx \frac{E(\mu,\sigma+h) - E(\mu,\sigma-h)}{2h}. \tag{3.34}$$

Implement this method for yourself and determine the minimal value of $E(\mu, \sigma)$. The choice of learning rate should not affect the result too much; in the example displayed in Fig. 3.8, $\gamma = 0.1$ is used.

When you have managed to minimize the expectation value of the energy – within reasonable precision, calculate the ground state energy by means of diagonalization and compare.

Finally, run this potential, Eq. (3.32), through your imaginary time procedure, the one you implemented in Exercise 3.4.3, and compare.

In Fig. 3.8 we show the $E(\mu, \sigma)$ landscape – along with a gradient descent sequence starting at $(\mu, \sigma) = (4, 2)$.

In these examples, where a direct solution of the time-independent Schrödinger equation by means of diagonalization is feasible, it does not make much sense to apply the variational principle. However, as we depart from the simple scenario of one particle in one dimension, the hope of obtaining the true ground state quickly becomes a pipe dream. For more realistic cases, the variational principle is actually a very useful one. Let's do just that and make such a departure.

[10] We don't have to use the same numerical increment h when calculating both the partial derivatives here, but it should still work if we do – as long as it is reasonably small.

3.4.5 Exercise: Variational Calculation in Two Dimensions

We minimize the energy of a potential that is an almost rectangular well in two dimensions:

$$V(x,y) = \frac{V_0}{(e^{s(|x|-w_x/2)}+1)(e^{s(|y|-w_y/2)}+1)}. \tag{3.35}$$

This potential, which is illustrated in Fig. 3.9, is the product of two one-dimensional potentials, of type $V_s(x)$, see Eq. (2.30). Here, we have set the smoothness, s, to be the same for both the x- and the y-part of the potential while the respective widths, w_x and w_y, differ. Specifically, take $V_0 = -1$, $s = 4$, $w_x = 4$ and $w_y = 2$. The corresponding Hamiltonian reads

$$\hat{H} = -\frac{\hbar^2}{2m}\frac{\partial^2}{\partial x^2} - \frac{\hbar^2}{2m}\frac{\partial^2}{\partial y^2} + V(x,y) = \hat{T}_x + \hat{T}_y + V(x,y), \tag{3.36}$$

where $\hbar^2/(2m) = 1/2$ in our usual units.

We use trial functions that are products of purely x- and y-dependent functions:

$$\psi_{\mathrm{trial}}(x,y) = \psi_x(x)\psi_y(y), \tag{3.37}$$

where $\psi_x(x)$ and $\psi_y(y)$ are normalized individually. The actual ground state wave function will not have this simple form. But that does not necessarily prevent our trial function from producing decent energy estimates anyway.

(a) Since the wave function now has two spatial variables, expectation values and other inner products now involve a double integral rather than the single integral of Eq. (1.24). Explain why the expectation value of the energy may now be estimated as

$$E = \langle\psi_x|\hat{T}_x|\psi_x\rangle + \langle\psi_y|\hat{T}_y|\psi_y\rangle + \int_{-\infty}^{\infty}\int_{-\infty}^{\infty} V(x,y)|\psi_x(x)\psi_y(y)|^2\,dx dy. \tag{3.38}$$

(b) Use your gradient descent implementation from Exercise 3.4.4, with suitable adjustments, to minimize the energy expectation value with Gaussian trial functions, $\psi_x(x) \sim \exp(-x^2/(4\sigma_x^2))$ and $\psi_y(y) \sim \exp(-y^2/(4\sigma_y^2))$ in Eq. (3.37). The

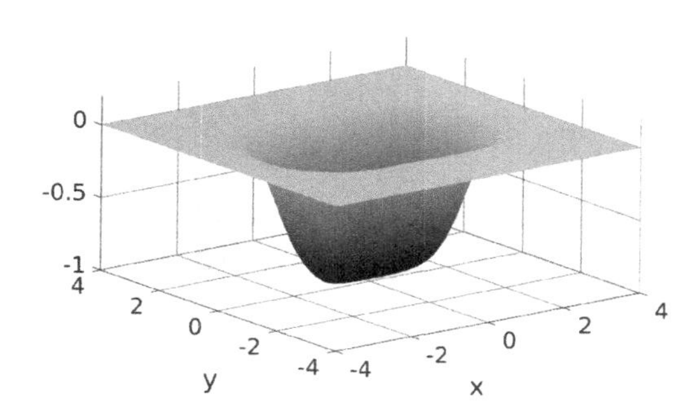

Figure 3.9 This bathtub-like shape is the potential in Eq. (3.35).

proper normalization factors are provided in Eq. (3.28). The parameters to be adjusted are σ_x and σ_y.

The double integral in Eq. (3.38) can be calculated summing over all combinations of x and y on their respective grids. This, in turn, could be implemented rather conveniently by first constructing matrices for both $V(x, y)$ and $|\psi(x, y)|^2$, then calculating their element-wise product, their *Hadamard product*, and summing over all elements. Finally, the double sum is multiplied[11] by h^2.

(c) Gaussians have a rather 'long tails', they fall off rather smoothly and never reach zero exactly. Perhaps, with a potential with rather sharp edges such as ours, trial functions that fall off a bit more abruptly would do better?

Repeat (b) with normalized cosine functions instead:

$$\psi_x(x) = \begin{cases} \sqrt{\frac{2a_x}{\pi}} \cos(a_x x), & -\frac{\pi}{2a_x} \leq x \leq \frac{\pi}{2a_x}, \\ 0, & \text{otherwise} \end{cases} \tag{3.39}$$

and correspondingly for ψ_y. Here, a_x and a_y are the parameters to tune in order to minimize E in Eq. (3.38).

Which guess gives the best estimate, the Gaussian or the cosine-shaped trial functions?

Does this conclusion seem to depend on the depth V_0; does it change if you redo the calculations with a different value for V_0?

In Fig. 3.10 we illustrate the optimal wave function for the case of a product of two Gaussians.

Now, these one-particle ground state estimates may have been complicated enough. However, as we addressed already in Section 1.3, things rapidly become more complex as the number of particles increases. There are, fortunately, ways of dealing with that.

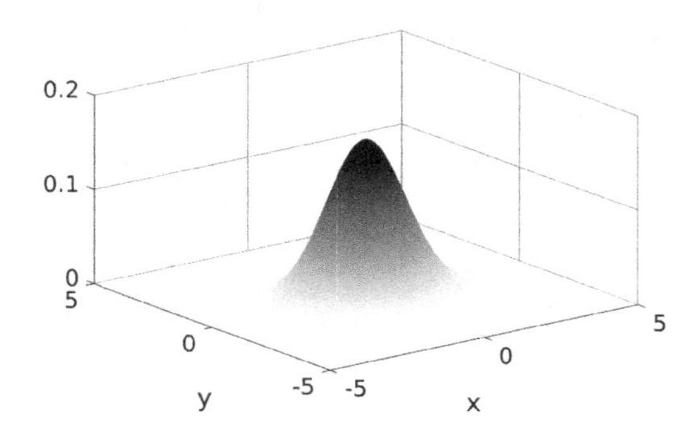

Figure 3.10 The *ansatz* with a product of two Gaussians in Eq. (3.37) produces a wave function that is far more smooth than the corresponding potential, Fig. 3.9. Still, it comes rather close to the actual ground state; the estimated energy is only 2% off.

[11] In case you, for some reason, choose to use separate grids for x and y, you should multiply by $h_x h_y$, not h^2.

4 Quantum Physics with Several Particles – and Spin

The 1977 Nobel laureate in Physics Philip W. Anderson made a very concise statement in the title of a *Science* publication: 'More is different' [4]. Although we have already seen several interesting phenomena with one particle alone, the quantum world becomes a lot richer, more complex and interesting when there are more particles. This is not only due to the infamous curse of dimensionality.

Contrary to Anderson, we will take 'more' to mean *two* in this chapter. Moreover, we will take these two particles to be two of the same kind; they will be *identical* – in the absolute sense.

4.1 Identical Particles and Spin

We can hardly talk about identical particles without addressing *spin*. It is a trait that quantum particles have. Just like a few other things we have encountered thus far, it does not have any classical analogue. It does, however, relate to angular momentum – although a simple picture of a ball rotating around its own axis would be a *too* simple one.

As mentioned in regard to Eq. (3.8), the 'normal' angular momentum $\boldsymbol{\ell}$, the angular momentum you also have in classical physics, is quantized in quantum physics. Also the spin $\mathbf{s}$ is quantized in a similar way; the eigenvalues of the square of the spin operator, $\hat{\mathbf{s}}^2$, are $s(s+1)\hbar^2$ – analogous to $\ell(\ell+1)h^2$ for $\hat{\boldsymbol{\ell}}^2$. But there are two significant differences.

(1) While the quantum number ℓ assumes integer values, the spin quantum number s may also assume half-integer values, $s = 0, 1/2, 1, 3/2, \ldots$

(2) While the angular momentum $\boldsymbol{\ell}$ of a system is a dynamical quantity – it may change due to interaction – the spin is an intrinsic property specific to the particle. A particle will never change its s quantum number.[1]

A particle with spin may, however, change the *direction* of its spin vector. Quantum particles that have a non-zero spin are affected by magnetic interactions. This was witnessed in the famous Stern–Gerlach experiment (see Fig. 4.1). It consisted in

[1] More precisely: an *elementary* particle would not change its spin; composite particles, such as an atom or a nucleus, may arrange the spins of the constituent particles together in various ways – leading to various *total* spins.

Figure 4.1 This plaque celebrates the beautiful experiment of Otto Stern (left) and Walther Gerlach (right). It is mounted on the building in Frankfurt, Germany, in which their famous 1922 experiment was conducted.

sending silver atoms through an inhomogeneous magnetic field. When the position of each atom was detected afterwards, it turned out that the atoms were deflected either upwards or downwards, with equal probabilities. Atoms passing straight through were not observed, they were always deflected up or down. This showed how the spin orientation is *quantized* – in addition to demonstrating the existence of *spin* in the first place. We will return to this experiment shortly.

You do not, however, need a magnetic field present in order for spin to be important. When we are dealing with two or more particles that are identical, keeping track of their spins is absolutely crucial.

4.1.1 Exercise: Exchange Symmetry

Suppose that $\Psi(x_1, x_2)$ is the wave function of a two-particle system that consists of identical particles. Let x_1 and x_2 be the positions of particles one and two, respectively.

(a) If the particles are truly identical in the absolute sense, any physical prediction must be independent of the order in which we list the particles.[2] As a consequence, we must insist that

$$\Psi(x_2, x_1) = e^{i\phi}\Psi(x_1, x_2) \tag{4.1}$$

for some real phase ϕ. Why?

[2] This is one of those ideas worth taking a moment to dwell on.

(b) Actually, this phase must be either 0 or π. In other words:

$$\Psi(x_2, x_1) = \pm\Psi(x_1, x_2). \tag{4.2}$$

Why?

Hint: Flip the ordering of the particles twice.[3]

So what's it going to be – plus or minus? The answer is: it depends. In this case it depends on the *spin* of the particles. Specifically, it depends on whether the spin quantum number s for the particles at hand is an integer or a half-integer. Electrons, for instance, have spin quantum number $s = 1/2$, while photons have spin with $s = 1$. Particles with half-integer spin, $1/2$, $3/2$ and so on, are called *fermions* while particles with integer spin, 0, 1, 2 ..., are referred to as *bosons*. The names are assigned in honour of the Italian physicist Enrico Fermi and the Indian mathematician and physicist Satyendra Nath Bose, respectively. And here is the answer to the question about the sign in Eq. (4.2):

The wave function for a set of identical fermions is anti-symmetric with respect to interchanging two particles, while wave functions for identical bosons are exchange symmetric.

In other words, Eq. (4.2) generalizes to any number of identical particles. This principle is called the *Pauli principle*, proposed by and named after the Austrian physicist Wolfgang Pauli, whose matrices we will learn to know quite well later on. We emphasize that this principle only applies to *identical* particles; the wave function of a system consisting of two particles of different types, such as an electron and a proton for instance, is not required to fulfil any particular exchange symmetry.

Here we have made several unsubstantiated claims about the spin of particles and their corresponding exchange symmetry. We are not able to provide any theoretical justification for these. The reason for this is the fact that they are fundamental postulates. The real justification is, as is so often the case with empirical sciences such as physics, the fact that it seems to agree very well with how nature behaves. The concept of spin grew out of a need to understand observation.

And we will keep on postulating a bit more. Like any angular momentum, spin is a vectorial quantity; it has a direction, not just a magnitude. The projection of the angular momentum, both the spin $\mathbf{s}$ and the 'usual' one $\boldsymbol{\ell}$, onto a specific axis is also quantized. If you measure the z-component s_z of the spin $\mathbf{s}$ of a particle, your result will necessarily be such that $s_z = m_s\hbar$, where $m_s = -s, -s+1, \ldots, s-1, s$. The case of $\boldsymbol{\ell}$ is quite analogous; the eigenvalues of $\hat{\ell}_z$ are $m_\ell\hbar$, where $m_\ell = -\ell, -\ell+1, \ldots, \ell-1, \ell$, see Eq. (3.8). For an $s = 1/2$ particle, only two values are possible for m_s, namely $-1/2$ and $+1/2$.

When we refer to the z-component of $\mathbf{s}$ and $\boldsymbol{\ell}$ above, this is just following the traditional labelling. The axis you project onto, regardless of what you call it, could point in any direction.

[3] In certain two-dimensional systems there are *quasiparticles* for which ϕ in Eq. (4.1) may actually take *any* value.

As mentioned, in the Stern–Gerlach experiment (Fig 4.1) silver atoms in their ground state were used. In this state the electrons organize themselves so that the total angular momentum is that of a single electron without 'ordinary' orbital angular momentum ℓ, so that, in effect, the whole atom becomes a spin-1/2 particle. Thus, a spin-projection measurement would reveal either $s_z = +1/2\,\hbar$ (up) or $s_z = -1/2\,\hbar$ (down) as result. When Walther Gerlach sent atoms through his magnetic field and then measured their positions afterwards, he performed precisely such a projection measurement; a particle with spin oriented upwards would be deflected upwards – and vice versa. The quantized nature of spin orientation is revealed by the fact that the silver atoms are deflected upwards or downwards at certain probabilities, never anything in between. If the atoms were considered classical magnets with arbitrary orientation in space, this distribution would be continuous – with a peak at $s_z = 0$. As demonstrated by the postcard that Walther Gerlach sent Niels Bohr in 1922 (see Fig. 4.2), this is not what came out of the experiment.

One more comment on the Stern–Gerlach experiment is in order. Sometimes textbooks have a tendency to present famous, fundamental experiments and their interpretations in overly coherent ways – ways that differ significantly from the original scope and the historical context of the experiment. The Davisson–Germer experiment, for instance, which consisted in bombarding a nickel surface with electrons at

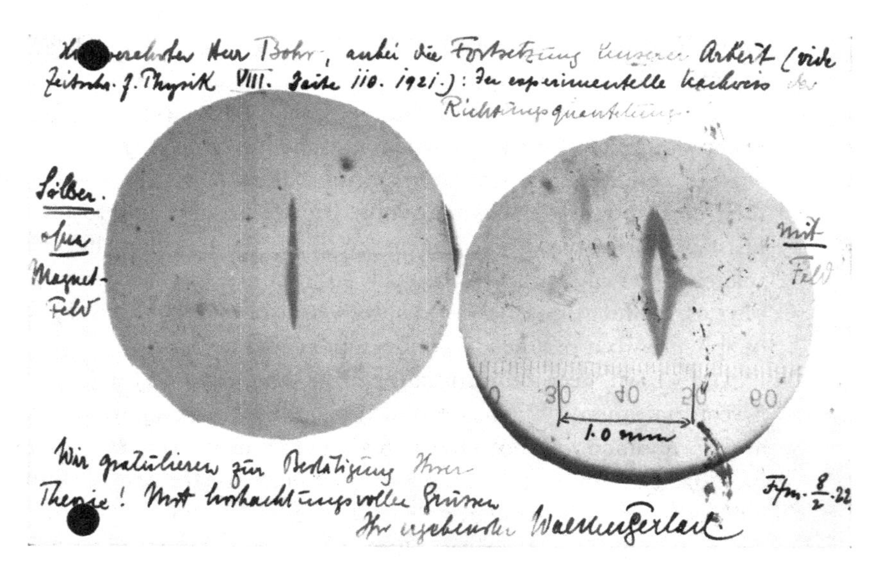

Figure 4.2 This postcard from Walther Gerlach to Niels Bohr, dated 8 February 1921, shows a photograph of the result of the famous Stern–Gerlach experiment, which Stern and Gerlach actually had agreed to give up on before Gerlach decided to give it one more go [18]. It says, 'Honourable Mr. Bohr, attached is the continuation of our work [. . .], the experimental proof of space quantization. [. . .] We congratulate you on the confirmation of your theory!' The left plate shows how the silver atoms are distributed in the absence of any magnetic field while the right one is the result with a field in place. The gap that emerges in the presence of a field shows how the atoms are deflected, either left or right in this case; no atom goes straight through.

different angles, is considered the first observation of quantum interference. However, the experimentalists simply set out to investigate the surface of nickel. When they ended up demonstrating that de Brogile was right about his wave assumption, this was more or less accidental. A similar comment may be made about the Michelson–Morley experiment and its relation to Einstein's special theory of relativity. The aim of Stern and Gerlach's experiment was to demonstrate the quantized nature of angular momentum of the atom, not *spin* angular momentum specifically. Actually, the notion of electron spin was not established until a few years later. This theoretical development involved names such as Ralph Kronig, George Uhlenbeck, Samuel Goudsmit and, albeit strongly opposed at first, Wolfgang Pauli.

4.1.2 Exercise: On Top of Each Other?

We have just learned that the spatial degree of freedom is not the only one that a quantum system has; *spin* also enters into the equation, so to speak. Thus, from now on, the wave function should also have a spin part.

But let us forget about that for one more minute and consider the spatial wave function of a system consisting of two identical fermions, $\Psi(x_1, x_2)$, which should abide by Eq. (4.2) with the proper sign.

(a) Suppose now that we measure the position of each fermion simultaneously. Why can we be absolutely certain that we will not get the same result for each particle? Can bosons be found on top of each other?

We next consider one spin-1/2 particle and construct its proper wave functions by augmenting the spatial part $\Psi(x)$ with a spin part. The general wave function may be written as a sum over the products of a spatial part and a spin part:

$$\Phi(x) = \Psi_\uparrow(x)\chi_\uparrow + \Psi_\downarrow(x)\chi_\downarrow, \tag{4.3}$$

where the χ are spin eigenstates corresponding to the spin quantum number $s = 1/2$; $\chi_\uparrow$ corresponds to $m_s = +1/2$ and $\chi_\downarrow$ to $m_s = -1/2$:

$$\hat{\mathbf{s}}^2\chi_\uparrow = \tfrac{1}{2}\left(\tfrac{1}{2}+1\right)\hbar^2\,\chi_\uparrow, \quad \hat{s}_z\chi_\uparrow = +\frac{1}{2}\hbar\,\chi_\uparrow, \tag{4.4a}$$

$$\hat{\mathbf{s}}^2\chi_\downarrow = \tfrac{1}{2}\left(\tfrac{1}{2}+1\right)\hbar^2\,\chi_\downarrow, \quad \hat{s}_z\chi_\downarrow = -\frac{1}{2}\hbar\,\chi_\downarrow. \tag{4.4b}$$

We insist that they are orthonormal according to their own inner product:

$$\langle\chi_\downarrow|\chi_\downarrow\rangle = \langle\chi_\uparrow|\chi_\uparrow\rangle = 1, \quad \langle\chi_\downarrow|\chi_\uparrow\rangle = 0. \tag{4.5}$$

When we are dealing with spin states, the notation $\langle\cdot|\cdot\rangle$ does not refer to the integral definition we used for spatial wave functions, Eq. (1.24). It is still a bona fide inner product, it's just a bit more abstract.[4]

If the spatial parts of the wave function in Eq. (4.3), $\Psi_\uparrow$ and $\Psi_\downarrow$, are proportional, the total wave function will be a single product state:

[4] In case you are thinking something along the lines of 'But what do these spin states actually look like?', don't worry, we will get to that. Let this be abstract for now.

$$\Phi(x) = \Psi(x)\chi, \tag{4.6}$$

where χ, the *spinor*, is a normalized linear combination of $\chi_\uparrow$ and $\chi_\downarrow$:

$$\chi = a\chi_\uparrow + b\chi_\downarrow, \quad \text{where} \quad |a|^2 + |b|^2 = 1. \tag{4.7}$$

The same applies to two-particle wave functions; they *could* be of the form

$$\Phi(x_1, x_2) = \Psi(x_1, x_2)\chi_2, \tag{4.8}$$

where χ_2 is a two-particle spin state. It can always be written as a linear combination of the four product states:

$$\chi_\uparrow^{(1)}\chi_\uparrow^{(2)}, \quad \chi_\uparrow^{(1)}\chi_\downarrow^{(2)}, \quad \chi_\downarrow^{(1)}\chi_\uparrow^{(2)} \quad \text{and} \quad \chi_\downarrow^{(1)}\chi_\downarrow^{(2)}. \tag{4.9}$$

In other words, a general two-particle spin state may be written

$$\chi_2 = a\, \chi_\uparrow^{(1)}\chi_\uparrow^{(2)} + b\, \chi_\uparrow^{(1)}\chi_\downarrow^{(2)} + c\, \chi_\downarrow^{(1)}\chi_\uparrow^{(2)} + d\, \chi_\downarrow^{(1)}\chi_\downarrow^{(2)} \tag{4.10}$$

for spin-1/2 particles. Here, the upper index refers to the particles in the same way that variables x_1 and x_2 refer to the positions of the respective particles. The first of these states, the ones listed in Eq. (4.9), corresponds to a state in which the spins of both particles are oriented upwards, the next one corresponds to particle 1 being orientated upwards while particle 2 is oriented downwards – and so on.

(b) Now, four specific linear combinations of spin states are particularly interesting:

$$\chi_\uparrow^{(1)}\chi_\uparrow^{(2)}, \quad \chi_\downarrow^{(1)}\chi_\downarrow^{(2)}, \quad \frac{1}{\sqrt{2}}\left(\chi_\uparrow^{(1)}\chi_\downarrow^{(2)} + \chi_\downarrow^{(1)}\chi_\uparrow^{(2)}\right), \tag{4.11a}$$

$$\frac{1}{\sqrt{2}}\left(\chi_\uparrow^{(1)}\chi_\downarrow^{(2)} - \chi_\downarrow^{(1)}\chi_\uparrow^{(2)}\right). \tag{4.11b}$$

What can we say about the exchange symmetry of each of these spin states? Or, in other words, if we interchange the labellings $1 \leftrightarrow 2$, does this affect the state? How?

(c) The principle that fermionic wave functions are anti-symmetric with respect to interchanging the particle labelling applies to the total, combined wave function Φ, not any spatial part or spin part separately. Now, suppose two identical particles are in a product state such as in Eq. (4.8), and that the two-particle spin state χ_2 coincides with the state of Eq. (4.11b). Explain why the spatial part $\Psi(x_1, x_2)$ must be exchange *symmetric* in this case.

(d) Can two fermions with a wave function as in (c) be found on top of each other? If so, what can be said about their spin orientations?

The states listed in Eqs. (4.11) are in fact eigenstates of the total spin operator for two spin-1/2 particles – the sum of the spin operators for each particle. Or, more precisely, they are eigenstates of the operator

$$\hat{\mathbf{S}}^2 = (\hat{\mathbf{s}}_1 + \hat{\mathbf{s}}_2)^2, \tag{4.12}$$

where $\hat{\mathbf{s}}_1$ and $\hat{\mathbf{s}}_1$ are the vectorial spin operators for each of the particles.[5] The eigenvalues of $\hat{\mathbf{S}}^2$ have the form $S(S+1)\hbar^2$. For the three states in Eq. (4.11a), the total spin

[5] For the sake of clarity, we have not explained why the states in Eqs. (4.11) actually are eigenstates of the $\hat{\mathbf{S}}^2$ operator; this is by no means supposed to be obvious.

quantum number $S = 1$. This group of three is usually referred to as the *spin triplet*. The state of Eq. (4.11b), which is called the *singlet* state, has $S = 0$.

One lesson to be learned from the above exercise is that two identical fermions can never be at the same position when they also have the same spin projections. Note that this applies irrespective of any mutual repulsion or attraction between the particles; it is more fundamental than that. More generally, two identical fermions can never be found in the same quantum state simultaneously. This is a rather common way of presenting the Pauli principle, which is, for this reason, often referred to the Pauli *exclusion* principle. This is a direct consequence of the fact that the wave function for a system of identical fermions changes sign when two particles are exchanged. It has profound implications. Suppose an atom consists of a nucleus and several electrons which, somehow, do not interact. Even in this odd case, in which the electrons would not repel each other via their electrostatic interaction, you would still never see any atom with all its electrons in the ground state. The Pauli principle forces each electron to populate a quantum state that is not already occupied – to pick a vacant seat on the bus, so to speak. Since electrons, in fact, do repel each other, the picture is more complex. But the Pauli principle is still inevitable in order for us to understand the structure of matter. If electrons, protons and neutrons were bosons, neither physics nor chemistry would resemble anything we know at all. Consequently, nor would anything that consists of matter.

The fact that identical bosons, on the other hand, are free to occupy the same quantum state has interesting consequences such as *superconductance* and *superfluidity*. We will, however, stick with fermions in the following.

4.2 Entanglement

The general shape of a single-particle wave function of a spin-1/2 particle in Eq. (4.3) is an example of what we called an *entangled* state. In general we cannot consider spatial and spin degrees of freedom separately. We can, however, if the wave function happens to be in a product form, such as in Eq. (4.6) or Eq. (4.8). For such situations, the space part and the spin part of the wave function are *disentangled*; it makes sense to consider one of them without regard for the other. In fact, this is what we did in the preceding chapters – before we introduced spin.

For a two-particle system of the form of Eq. (4.8), the space part Ψ or the spin part χ_2 could also be subject to entanglement individually. As an example, let us again consider the four spin states in Eq. (4.9) and in Eqs. (4.11). If our state is prepared in, say, the third product state of Eq. (4.9), we know for sure that particle 1 has its spin oriented downwards while particle 2's spin is oriented upwards. The same may not be said about the third state in Eqs. (4.11); particle 1 could have its spin oriented either way – and so could particle 2. However, they are not independent, this two-particle spin state only has components in which the spin projections are opposite. The state is rigged such that a spin measurement on one of the particles will also be decisive for the other. Suppose Walther Gerlach, or someone else, measures the spin of particle 1. That would induce a collapse of the spin state of particle 1 to one of the $\hat{s}_z$ eigenstates,

$\chi_\uparrow^{(1)}$ or $\chi_\downarrow^{(1)}$. If Walther finds the spin to point downwards, the resulting two-particle spin state is collapsed to $\chi_\downarrow^{(1)}\chi_\uparrow^{(2)}$; in our initial state the component d in Eq. (4.10) is zero and, thus, so is the probability of reading off spin down for particle 2.

So, Walther does not have to measure the spin of particle 2 to know the outcome, it has to be oriented upwards. And, while the outcome of the first measurement could just as well have been upwards, a measurement on the second one would still have yielded the opposite result. In other words, if a spin-1/2 system with two particles is prepared in the third state of Eqs. (4.11a), we do not know anything about the spin orientation of each of the particles individually, but we may be certain that they are oriented oppositely. The same may also be said about the state in Eq. (4.11b), the singlet state.

4.2.1 Exercise: Entangled or Not Entangled

When full knowledge about the system boils down to full knowledge of each particle, the state of a quantum system with two particles is described by a product state. When this is not possible, we say that the particles are *entangled*.

(a) Below we list five different normalized states for a system of two spin-1/2 particles. Which of these are entangled and which are not?

(i) $\frac{1}{\sqrt{2}}\left(\chi_\downarrow^{(1)}\chi_\downarrow^{(2)} - \chi_\downarrow^{(1)}\chi_\uparrow^{(2)}\right)$

(ii) $\frac{1}{\sqrt{2}}\left(\chi_\downarrow^{(1)}\chi_\downarrow^{(2)} - \chi_\uparrow^{(1)}\chi_\uparrow^{(2)}\right)$

(iii) $\chi_\uparrow^{(1)}\chi_\downarrow^{(2)}$

(iv) $\frac{1}{2}\left(\chi_\uparrow^{(1)}\chi_\uparrow^{(2)} + \chi_\uparrow^{(1)}\chi_\downarrow^{(2)} - \chi_\downarrow^{(1)}\chi_\uparrow^{(2)} - \chi_\downarrow^{(1)}\chi_\downarrow^{(2)}\right)$

(v) $\frac{1}{\sqrt{3}}\left(\chi_\uparrow^{(1)}\chi_\uparrow^{(2)} + \chi_\uparrow^{(1)}\chi_\downarrow^{(2)} - i\,\chi_\downarrow^{(1)}\chi_\downarrow^{(2)}\right)$

(b) We claimed that in order for the total wave function including both the spatial and spin degrees of freedom, Eq. (4.3), to be on product form, Eq. (4.6), $\Psi_\uparrow(x)$ and $\Psi_\downarrow(x)$ would have to be proportional. Otherwise space and spin would feature some degree of entanglement. Show this.

(c) In Eqs. (4.11) we saw two entangled states and two product states. The former states, the entangled ones, correspond to a situation in which the spin projection of each particle is entirely undetermined, while we know they are oppositely aligned. Can you construct linear combinations of the product states in Eqs. (4.11a) for which the same may be said – except with aligned spins?[6]

(d) Suppose that you pick coefficients in the linear combination of Eq. (4.10) at random. It this likely to amount to an entangled state? Or do you have to carefully design your linear combination in order for entanglement to emerge?

[6] You may have seen one such state already.

We now touch upon another profound characteristic of nature, namely the *non-locality of quantum physics*. We have learned that if a two-particle system is prepared in a fully entangled state such as the singlet state of Eqs. (4.11b), knowledge of the outcome of a measurement on the first particle will determine the outcome of a measurement on the second – unless it has been tampered with. This is supposed to hold irrespective of the spatial separation between the two particles.

But wait. How, then, can the second particle have learned which way to orient its spin after a measurement on the first one? Surely, the first particle cannot possibly have sent the second one any instantaneous message – defying the speed of light, not to mention common sense. In reality, the orientations must have been rigged before we moved the particles apart, right? Although we were not able to predict the outcome of the first measurement, that couldn't mean that it was actually arbitrary; there must have been a piece of information we missed out on. Right?

This issue is one of the reasons why Einstein never fully bought into quantum physics. It relates to one of the most famous disputes in science, in which Einstein and Bohr played the opposing parts (Fig. 4.3). Bohr argued that particle correlations such as the ones we have studied here could very well persist in spite of arbitrarily large spatial separation; no communication between the particle was needed.

Figure 4.3 In 1925, their mutual friend Paul Ehrenfest, himself a brilliant physicist, brought Niels Bohr and Albert Einstein together in an attempt to have them settle their scientific differences about the interpretation of quantum theory. Despite the seemingly good atmosphere, he did not succeed.

Many physicists would say that the matter was settled when experiments ingeniously devised by John Stewart Bell, from Northern Ireland, were carried out in the early 1970s. States such as the last two in Eqs. (4.11) were crucial in this regard. In fact, these are examples of so-called *Bell states*.

And who was right, you wonder? The short answer is that Bohr was right and Einstein was wrong.

The concept of entanglement is certainly apt for philosophical discussions. It also has very specific technological implications when it comes to quantum information technologies, some of which will be addressed in the next chapter.

In most of the remainder of *this* chapter, we will see how spin, via the exchange symmetry, strongly affects the energy of two-particle states. Before investigating this further, however, we will dwell a little more on the spin characteristics alone.

4.3 The Pauli Matrices

Perhaps it seems odd to consider the spin part without any regard to the spatial dependence of the wave function. However, this does makes sense if we are able to pin the particles down somehow – for instance in an ion trap, as in Fig. 4.4.

If we take the states $\chi_\uparrow$ and $\chi_\downarrow$ to be the basis states, we may conveniently represent the general spin state of a spin-1/2 particle as a simple two-component vector:

$$\chi = a\chi_\uparrow + b\chi_\downarrow \to \begin{pmatrix} a \\ b \end{pmatrix}. \tag{4.13}$$

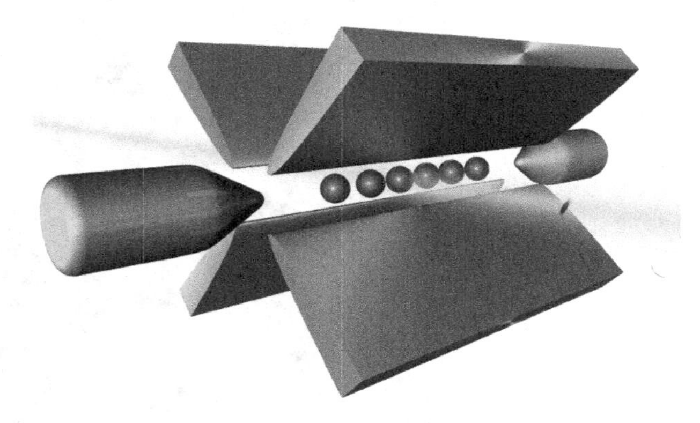

Figure 4.4 Illustration of an ion trap. An ion differs from an atom in that the number of electrons does not match the number of protons in the nucleus. Thus, contrary to atoms, ions carry a net charge. In an ion trap, ions are confined in space by electric fields. Laser beams may now be used to manipulate and control the internal state of individual ions – or a few neighbouring ions.

Correspondingly, inner products are conveniently calculated as

$$\langle \chi_1 | \chi_2 \rangle = \chi_1^\dagger \, \chi_2 = \begin{pmatrix} a_1^* & b_1^* \end{pmatrix} \begin{pmatrix} a_2 \\ b_2 \end{pmatrix} = a_1^* a_2 + b_1^* b_2. \tag{4.14}$$

The spin orientation of a particle can be manipulated by magnetic fields. Specifically, if a particle with spin is exposed to a magnetic field $\mathbf{B}$, this adds the contribution

$$H_\mathrm{B} = -g \frac{q}{2m} \mathbf{B} \cdot \mathbf{s} \tag{4.15}$$

to the Hamiltonian. Here q is the charge of the particle and g is a somewhat mysterious factor which emerges in relativistic formulations of quantum physics. For an electron, this g-factor happens to be 2, or, more precisely, very close to 2. As any physical quantity, the spin has its own operator. In the representation of Eq. (4.13), in which $\chi_\uparrow = (1,0)^\mathrm{T}$ and $\chi_\downarrow = (0,1)^\mathrm{T}$, the vectorial spin operator may be expressed in terms of matrices:

$$\hat{\mathbf{s}} = [\hat{s}_x, \hat{s}_y, \hat{s}_z] = \frac{\hbar}{2} \left[\sigma_x, \sigma_y, \sigma_z \right] = \frac{\hbar}{2} \boldsymbol{\sigma}, \tag{4.16}$$

where

$$\sigma_x = \begin{pmatrix} 0 & 1 \\ 1 & 0 \end{pmatrix}, \quad \sigma_y = \begin{pmatrix} 0 & -\mathrm{i} \\ \mathrm{i} & 0 \end{pmatrix}, \quad \sigma_z = \begin{pmatrix} 1 & 0 \\ 0 & -1 \end{pmatrix}. \tag{4.17}$$

These are the famous *Pauli matrices*.

4.3.1 Exercise: Some Characteristics of the Pauli Matrices

(a) Verify, by direct inspection, that $\sigma_x^2 = \sigma_y^2 = \sigma_z^2 = I_2$, where I_2 is the 2×2 identity matrix.

(b) Verify that *any* spin-1/2-particle state, Eq. (4.13), is an eigenstate of $\hat{\mathbf{s}}^2 = \hat{s}_x^2 + \hat{s}_y^2 + \hat{s}_z^2$ with eigenvalue $s(s+1)\hbar^2$, $s = 1/2$. See Eqs. (4.4).

(c) Verify that the Pauli matrices are all Hermitian.

(d) Verify that $\left[\sigma_x, \sigma_y\right] = 2\mathrm{i}\sigma_z$. We introduced the commutator between two operators, or matrices, in Eq. (2.42). The same holds if you interchange x, y and z cyclically, $\{x, y, z\} \to \{y, z, x\} \to \{z, x, y\}$.

(e) Verify that $\left\{\sigma_x, \sigma_y\right\} = \{\sigma_x, \sigma_z\} = \left\{\sigma_y, \sigma_z\right\} = 0_{2,2}$, the 2×2 zero matrix, where

$$\{A, B\} = AB + BA \tag{4.18}$$

is the *anti-commutator*.

As mentioned, the basis states of our spin space, $\chi_\uparrow$ and $\chi_\downarrow$, are the eigenstates of the spin projection along the z-axis. For a particle prepared in the $\chi_\uparrow$ state, a measurement of its spin along the z-axis will necessarily give the eigenvalue $+1/2\,\hbar$ of the operator $\hat{s}_z = \hbar/2\,\sigma_z$. As also mentioned, the z-axis is not by any means more special than any other axis, and it could be chosen to point in any direction. Moreover, we may very well want to measure the spin projection along the x- or the y-axes as well.

4.3.2 Exercise: The Eigenstates of the Pauli Matrices

(a) It is quite straightforward to observe that for σ_z the eigenvalues of the eigenvectors $\chi_\uparrow = (1,0)^{\mathrm{T}}$ and $\chi_\downarrow = (0,1)^{\mathrm{T}}$ are $+1$ and -1, respectively – corresponding to the eigenvalues $\pm\hbar/2$ for $\hat{s}_z$ in Eqs. (4.4).

Show that the eigenvalues of σ_x and σ_y are also ± 1. What are the corresponding eigenvectors?

(b) If you measure the spin projection along the z-axis and find that it is $1/2\,\hbar$, what will be the outcome of performing the same measurement once more right after?

(c) Suppose now that the spin projection along the x-axis has been measured and found to be positive – it turned out to be $+1/2\,\hbar$. This measurement collapsed the spinor into the corresponding eigenvector; let's call it $\chi_\rightarrow$. Next, we make another spin projection measurement, this time along the z-axis. What is the probability of measuring 'spin up' – of finding $+1/2\,\hbar$ for s_z? And what is the probability of measuring spin down?

(d) Now, let's flip the coin and assume that we have measured a positive spin projection along the z-axis. What is the probability of measuring a positive spin projection along the x-axis?

So, a sequence of spin projection measurements where you constantly change between the z-axis and the x-axis is like tossing a coin, in the sense that the result could, with equal probability, be either positive or negative every time. However, if you do repeated measurements along the same axis, you are bound to get the same result as the first time – over and over again. These are direct consequences of how the wave function, also the spin wave function, collapses due to a measurement.

4.4 Slater Determinants, Permanents and Energy Estimates

For many-particle systems, it is quite common when solving both the time-dependent and the time-independent Schrödinger equation to express the solution in terms of single-particle wave functions. Typically, these wave functions are eigenfunctions of a single-particle Hamiltonian; for one-particle systems we saw examples of this in Section 3.3. Suppose now that we have a complete set of one-particle wave functions $\varphi_1(x), \varphi_2(x), \ldots$, which include both a spin part and a spatial part. We may use these to build many-particle states – to construct a basis in which we could expand our many-body wave function. The simplest way to construct such a basis would be to simply put together all possible products of type

$$\Phi(x_1, x_2, x_3, \dots) = \varphi_1(x_1)\varphi_2(x_2)\varphi_3(x_3)\dots. \tag{4.19}$$

The actual wave function can always be written as a linear combination of such product states. However, if these particles are identical fermions, we must make sure that this linear combination indeed constitutes an exchange anti-symmetric wave function. This could potentially be rather cumbersome.

If, instead of such simple product states, we construct a basis of N-particle wave functions where each of them already has the proper exchange symmetry, this would no longer be an issue; any linear combination of these states would then also have the right exchange symmetry. For identical fermions, a very convenient way of achieving this is to write our basis states as *determinants*:

$$\Phi(x_1, x_2, \dots, x_N) = \frac{1}{\sqrt{N!}} \begin{vmatrix} \varphi_1(x_1) & \varphi_1(x_2) & \cdots & \varphi_1(x_N) \\ \varphi_2(x_1) & \varphi_2(x_2) & \cdots & \varphi_2(x_N) \\ \vdots & \vdots & \ddots & \vdots \\ \varphi_N(x_1) & \varphi_N(x_2) & \cdots & \varphi_N(x_N) \end{vmatrix}. \tag{4.20}$$

Such determinants are called *Slater determinants* after the US physicist John C. Slater who introduced the concept in 1929. The prefactor $1/\sqrt{N!}$ ensures normalization – provided that the single-particle states $\varphi_1(x), \varphi_2(x), \dots$ are orthonormal. Here we have constructed an N-particle state consisting of the N first states in the single-particle basis. If one or more of the single-particle wave functions is replaced by others, we would produce another, orthonormal N-particle wave function. Rearranging the ones we already have would not, however. As you may remember from your classes in linear algebra, interchanging two columns – or rows – in a quadratic matrix only amounts to an overall sign change in its determinant, thus ensuring exchange anti-symmetry.

So much for fermions, what about bosons? Can we construct a basis of exchange symmetric N-particle states from the same single-particle states? Sure we can. It's simple; just replace the determinant in Eq. (4.20) with its less known sibling, the *permanent*. The determinant of a square matrix is defined as the sum of all possible signed products of matrix entries that share neither row nor column. The sign is determined by the order of the element indices. The permanent is defined in the same way – except there is no sign rule. Consequently, exchanging two columns does not change anything in the permanent; a Slater permanent constructed analogously to Eq. (4.20) will, correspondingly, be exchange symmetric by construction.

4.4.1 Exercise: Repeated States in a Slater Determinant

It does not make sense to include the same single-particle state more than once in a Slater determinant. Why is that?

Does it make sense to include repeated single-particle states in a Slater *permanent*?

The answers to these questions are the underlying reasons why ground states of bound two-particle systems, such as the helium atom, are spatially exchange-symmetric states, not anti-symmetric ones.

4.4.2 Exercise: The Singlet and Triplet Revisited

Once again we only consider the spin part of a particle – using $\chi_\uparrow$ and $\chi_\downarrow$ as our basis states. For a system with $N = 2$ spin-1/2 particles, show that the only non-zero Slater determinant is in fact the singlet state in Eq. (4.11b).

Suppose we use the same basis states to construct two-particle Slater *permanents*. Which states will we arrive at then?

Approaches in which the wave function is expressed as a linear combination of determinants/permanents are frequently used in solving many-body problems – predominantly time-independent ones. The most straightforward approach in this regard is called *full configuration interaction*. In this method, one starts out with a set of orthonormal single-particle states, one that hopefully is near complete, construct all possible Slater determinants/permanents from this set, construct the Hamiltonian matrix in this basis and, finally, diagonalize it.

While this may be intuitively appealing, it is usually rather costly computationally – too costly in most interesting cases. This is rarely a viable way to battle the curse of dimensionality. Instead we could try to make educated guesses on what the wave function might look like – assume a certain shape of the wave function – and, in this way, reduce complexity considerably. When making such 'guesses' for the wave function, which depends on both positions and spins, we must be sure to encompass the proper exchange symmetry.

4.4.3 Exercise: Variational Calculations with Two Particles

As we did in several exercises in Chapter 3, we make use of the variational principle, Eq. (3.27), to estimate ground state energies. Contrary to those examples, here we consider a two-particle system.

Suppose two identical fermions are trapped in a confining potential of the form of Eq. (2.30), and that they interact via this potential:

$$W(x_1, x_2) = \frac{W_0}{\sqrt{(x_1 - x_2)^2 + 1}}. \tag{4.21}$$

Here, as before, x_1 is the position of the first particle and x_2 is the position of the second. We take the interaction strength W_0 to be 1 in this case. With positive W_0, the interaction is a repulsive one; the energy increases as the two particles approach each other – when $|x_1 - x_2|$ becomes small. Our interaction could model the electric repulsion between two particle with charges of the same sign.[7]

[7] Since we work with one-dimensional particles, it would not work to use the actual Coulomb potential, Eq. (2.7), in this case.

With two particles, our spatial wave function has two variables, $\Psi(x_1, x_2)$, and the Hamiltonian is a two-particle operator:

$$\hat{H} = \hat{h}_1 + \hat{h}_2 + W(x_1, x_2), \tag{4.22}$$

where $\hat{h}_k = \hat{p}_k^2/(2m) + V(x_k)$ is the one-particle part[8] for particle number k. The energy expectation value is now a double integral:

$$\langle E \rangle = \langle \Psi | \hat{H} | \Psi \rangle = \int_{-\infty}^{\infty} \int_{-\infty}^{\infty} [\Psi(x_1, x_2)]^* \; \hat{H} \, \Psi(x_1, x_2) \, \mathrm{d}x_1 \, \mathrm{d}x_2, \tag{4.23}$$

which has three terms – corresponding to the three terms of the Hamiltonian, Eq. (4.22).

And then there is spin.

(a) First, assume that the two-particle system has a spatial wave function which can be written as a product of the same one-particle wave function for each of the particles:

$$\Psi(x_1, x_2) = \psi_{\mathrm{trial}}(x_1) \, \psi_{\mathrm{trial}}(x_2). \tag{4.24}$$

This is a crude oversimplification. However, it may still be able to produce a decent energy estimate.[9]

In terms of exchange symmetry, is such a spatial wave function admissible for fermions? If we include the spin degree of freedom in the wave function, could Eq. (4.24) be part of a Slater determinant? What requirements apply to the spin state of these two particles in that case?

(b) Assume that the spatial trial wave function ψ_{trial} has the form of Eq. (3.28), a Gaussian centred around $x = 0$. It depends parametrically on the width σ. Now, as in Exercise 3.4.2, minimize this expectation value with respect to this parameter. Let your confining potential, $V(x)$, be given by Eq. (2.30), with the parameters $V_0 = -1$, $w = 4$ and $s = 5$.

As in Exercise 3.4.5, this is a bit more cumbersome than many other exercises we have seen since we are dealing with double integrals. However, similar to Eq. (3.38), the two terms stemming from the one-particle Hamiltonians, $\hat{h}_1$ and $\hat{h}_2$, can be reduced to single integrals – even identical ones in this case. The interaction energy, on the other hand, remains a double integral – one that you can deal with in exactly the same manner as in Exercise 3.4.5.

As always: Ensure that your grid is dense and wide enough by repeating your calculations with improved numerics.

(c) Next, allow our trial function a bit more of 'wiggle room' and base our one-particle spatial wave function on ψ_{Gauss} of Eq. (3.31) from Exercise 3.4.4. This trial function is still a Gaussian; however, this one depends on an offset parameter μ in

[8] Strictly speaking, since our wave function resides in a function space corresponding to *two* particles, it is more correct to write the one-particle part for particle 1 as the tensor product $\hat{h} \otimes \hat{I}$ and $\hat{I} \otimes \hat{h}$ for particle 2, where $\hat{I}$ is the identity operator.

[9] We saw in Exercise 3.4.5 that such a simple product could work rather well for one particle in two dimensions. The case of two particles in one dimension is, of course, quite different, though.

addition to σ. We let the offset of one particle be the negative of the other. This could motivate a spatial wave function of this form:

$$\Psi(x_1, x_2) = \psi_{\text{Gauss}}(x_1;\sigma, \mu)\, \psi_{\text{Gauss}}(x_2;\sigma, -\mu). \tag{4.25}$$

Why is this one not admissible? [10]

And why is this one OK?

$$\Psi(x_1, x_2) = N\big[(\psi_{\text{Gauss}}(x_1;\sigma, \mu)\, \psi_{\text{Gauss}}(x_2;\sigma, -\mu) \\ + \psi_{\text{Gauss}}(x_1;\sigma, -\mu)\, \psi_{\text{Gauss}}(x_2;\sigma, \mu)\big]. \tag{4.26}$$

Is this a Slater determinant – or a permanent?

Here the factor N is imposed in order to ensure normalization; although ψ_{Gauss} is normalized, $\psi_{\text{Gauss}}(x;\sigma, \mu)$ and $\psi_{\text{Gauss}}(x;\sigma, -\mu)$ will in general overlap; they only become orthogonal in the limit $\mu \to \infty$.

(d) Actually, we may ignore the normalization factor N if we formulate the variational principle, Eq. (3.27), slightly differently:

$$\langle E \rangle = \frac{\langle \Psi | \hat{H} | \Psi \rangle}{\langle \Psi | \Psi \rangle} \geq \varepsilon_0. \tag{4.27}$$

Now, again minimize the energy expectation value with the trial wave function of Eq. (4.26). This time we are dealing with two parameters, σ and μ. Your gradient descent implementation from Exercise 3.4.4 may come in handy.

(e) Thus far we have been dealing with spatial wave functions that are exchange symmetric. This is admissible for two-particle wave functions of the product form of Eq. (4.8) where the spin part χ_2 is the spin single state, Eq. (4.11b). However, we could choose χ_2 to be one of the triplets – with the consequence that the spatial part must be exchange anti-symmetric:

$$\Psi(x_1, x_2) = N\big[\psi_{\text{Gauss}}(x_1;\sigma, \mu)\psi_{\text{Gauss}}(x_2;\sigma, -\mu) \\ - \psi_{\text{Gauss}}(x_1;\sigma, -\mu)\psi_{\text{Gauss}}(x_2;\sigma, \mu)\big]. \tag{4.28}$$

Now, repeat the minimization from (d) with this wave function instead. How does going from an exchange symmetric to an anti-symmetric spatial wave function affect the energy estimate?

In Fig. 4.5 we display the wave function approximations corresponding to the two first spatially exchange-symmetric assumptions in Exercise 4.4.3 – the ones that minimize the energy expectation value. We also plot the actual ground state wave function – along with the result of the approach we will in address in Exercise 4.4.4. Due to the mutual repulsion, Eq. (4.21), the two particles tend to be a bit separated. This is clearly seen in Fig. 4.5(b) and (d), and more so in (d). The simple product assumption of Eq. (4.24) is not able to encompass this at all; in fact, it predicts a rather high probability for finding the two particles on top of each other ($x_1 = x_2$). Despite this shortcoming, and the crude assumption in Eq. (4.24), the energy prediction is not that bad, it's only

[10] Assuming that $\mu \neq 0$, that is.

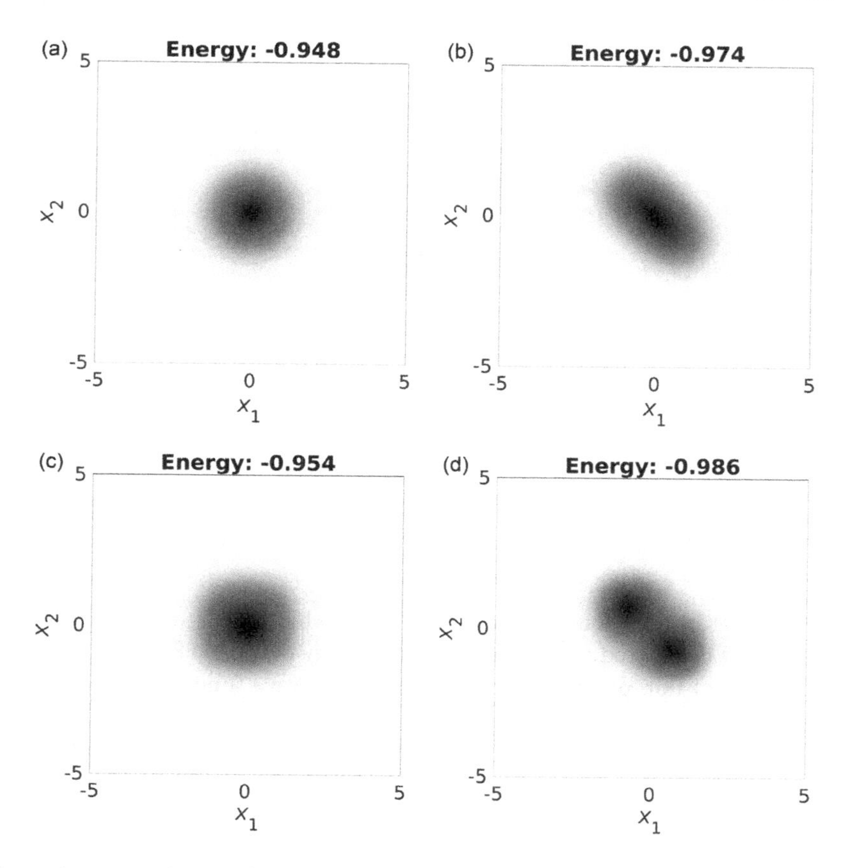

Figure 4.5 Plots of ground state approximations for the problem in Exercise 4.4.3. All of them correspond to wave functions with an exchange-symmetric spatial part. The corresponding energy estimates are written above each of them. As we see, they all produce rather similar energy estimates, although the shape of the wave function estimates differ quite a bit. (a) The product state in Eq. (4.24), and (b) the spatially exchange-symmetric assumption in Eq. (4.26); (c) the approach of Exercise 4.4.4; (d) the actual ground state wave function.

off by 4%. The assumption of Eq. (4.26) does allow the particles to be apart, and, correspondingly, the energy estimate is better. The energy estimate of 0.974 energy units is only 1.2% too high.

Actually, this is rather typical – the ground state energy estimate is surprisingly good although the trial wave function looks a bit off. Is there any particular reason why it should be so? Should the energy expectation value be less sensitive to adjustments of the wave function for the ground state than any other state? What does the variational principle say on the matter?

Both the assumptions of Eqs. (4.24) and (4.26) restrict the wave functions to consist of Gaussians. In Exercise 4.4.4 we will lift this restriction.

But before we dedicate ourselves to that task, let's have a look at the exchange anti-symmetric assumption of Eq. (4.28). The wave function with σ and μ values that minimize the energy expectation value is shown in Fig. 4.6. It estimates an energy,

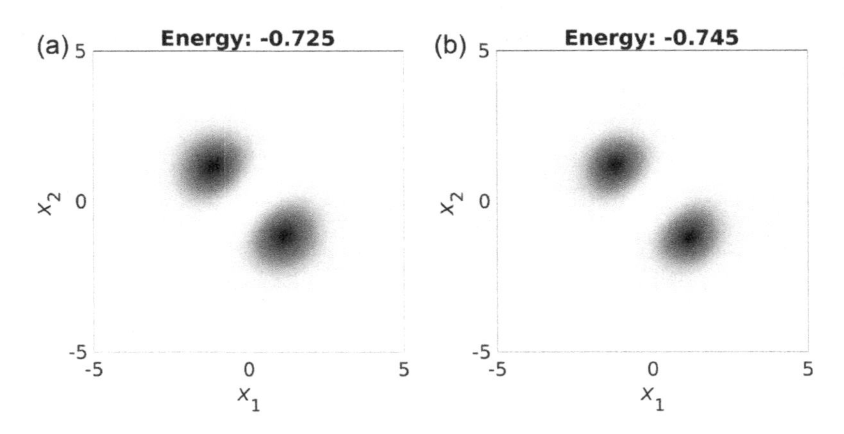

Figure 4.6 (a) The absolute square of the wave function that minimizes the energy assuming the spatial form of Eq. (4.28). (b) The actual minimal energy eigenstate which is exchange anti-symmetric. Note that in all cases, due to the anti-symmetry, the wave function is zero whenever $x_1 = x_2$, in accordance with what we saw in Exercise 4.1.2.

or rather, sets an upper bound for it, which reads -0.7235 energy units, which is way off compared to the ground state energy of -0.9858 energy units. However, this is not a fair comparison; our anti-symmetric wave function is not really an estimate of the ground state.

The exact ground state is in fact the product of a spin and a spatial part, Eq. (4.8), with a *symmetric* spatial part – combined with the spin singlet. The spatially anti-symmetric estimate in Fig. 4.6(a) is actually an estimate of a spin-symmetric (triplet) excited state – the one with the lowest energy within this symmetry. The true excited state – with a fairly accurate numerical energy estimate – is displayed in Fig, 4.6(b). We believe it is fair to say that the estimate based on Eq. (4.28) does rather well after all.

This goes to show how important spin is – also for the energy structure of quantum systems with no spin dependence whatsoever in their Hamiltonians. The spin degree of freedom simply cannot be ignored for identical particles, due to the requirement on exchange symmetry. This, in turn, is quite decisive for the energy that comes out.

Now, as promised, let's get rid of the restriction that the single-particle trial functions have to be Gaussians.

4.4.4 Exercise: Self-Consistent Field

We revisit the problem of Exercise 4.4.3. We again assume that the ground state can be written as a product state with two identical wave functions,[11] as in Eq. (4.24). But instead of fixing the shape of the one-particle wave function as a Gaussian, or any other specific form for that matter, we here determine it in an iterative manner.

First, we fix an initial guess for a one-particle trial function $\psi_{\text{trial}}(x_1)$ for the first particle. Next we assume that the other particle 'feels' the first one via a static charge

[11] At the risk of being excessively explicit about this, it can't. We can only hope that this is a decent approximation.

distribution set up by its wave function. This motivates an additional one-particle potential felt by particle 2, set up by particle 1:

$$V_{\text{eff}}(x_2) = \int_{-\infty}^{\infty} |\psi_{\text{trial}}(x_1)|^2 W(x_1, x_2)\, dx_1, \tag{4.29}$$

where the interaction W is given in Eq. (4.21).[12] In this scenario, the second particle will also have the same effect on the first particle. With this, we can set up a one-particle Hamiltonian for each particle with this addition, diagonalize it and replace our single-particle wave function ψ_{trial} with the normalized ground state of this effective one-particle system. In the particularly simple product state *ansatz*, Eq. (4.24), there is only one single-particle trial function in play at all times.

In other words, we set out to solve the effective one-particle Schrödinger equation:

$$\left[\hat{h} + V_{\text{eff}}(x)\right] \psi(x) = \varepsilon\, \psi(x). \tag{4.30}$$

Note that this Schrödinger-like equation, contrary to all other examples we have seen thus far, is not actually a linear differential equation anymore since V_{eff} in fact depends on ψ itself.

So how do we go about finding a ψ that produces an effective potential which, in turn, renders itself as the ground state of the effective Hamiltonian? The answer is: we iterate.

We start off with some educated guess for ψ_{trial} and determine the effective potential set up by this one-particle wave function, Eq. (4.29). Next, we construct the effective Hamiltonian, diagonalize it and obtain a new ψ_{trial} by picking the ground state of this effective one-particle Hamiltonian. In going from one iteration to the next, we replace one ψ_{trial} with another, which leads to an altered effective potential which, in turn, alters ψ_{trial} and so on. We repeat this iteration procedure until ψ_{trial} does not really change anymore – until the system is *self consistent*.

Finally, when we are pleased with our ψ_{trial}, the energy expectation value of our two-particle trial function may, as before, be calculated as

$$\langle E \rangle = \langle \Psi | \left(\hat{h}_1 + \hat{h}_2 + W\right) | \Psi \rangle, \tag{4.31}$$

where $\Psi(x_1, x_2) = \psi_{\text{trial}}(x_1)\, \psi_{\text{trial}}(x_2)$. For the sake of clarity: the one-particle parts of the above Hamiltonian, $\hat{h}_1$ and $\hat{h}_2$, which, respectively, affect particles 1 and 2 exclusively, do not contain the effective potential in Eq. (4.29) – these are the actual one-particle Hamiltonians.[13]

Implement this iteration procedure and use it to estimate the two-particle ground state energy. As your starting choice for the one-particle wave function ψ_{trial} you may very well make our usual choice and take it to be a normalized Gaussian centred around $x = 0$, Eq. (3.28). In our usual grid representation, the effective potential V_{eff} is represented as a diagonal matrix – just like the 'ordinary' potential V. For each diagonal element of the V_{eff}-matrix, you must calculate an integral. You can monitor the

[12] The particles' charges enter via the parameter W_0 in Eq. (4.21).

[13] Having said this, we should also say that the expression in Eq. (4.31) *could* be written rather compactly using V_{eff}.

convergence by checking the lowest eigenvalue in Eq. (4.30) for each iteration; when it no longer changes much, we may stop iterating and calculate our energy estimate, Eq. (4.31).

Is the energy estimate you found lower or higher than the estimate you found in Exercise 4.4.3(b)? Does this alone enable you to say which estimate is the best?

The approach of Exercise 4.4.4 is a simple example of a *Hartree–Fock* calculation. Above we made a rather sketchy argument in which we assumed single-particle wave functions to set up static charge distributions. However, the method has a much more solid theoretical foundation: *It is the closest approximation we can have to the ground state energy under the assumption that the state is a single Slater determinant with orthonormal single-particle wave functions.*

The original method proposed by Douglas Hartree was even simpler; it minimized the energy under the assumption that the wave function was a product state, Eq. (4.19). Vladimir Fock's contribution was to restore of lack of exchange anti-symmetry by taking a Slater determinant, Eq. (4.20), as the starting point instead. This, in turn, introduces an additional effective potential which is referred to as the *exchange potential*. Contrary to the effective potential of Eq. (4.29), which is called the *Hartree potential* or the *direct potential*, it is not local in position; a matrix representation of the exchange term on a numerical grid would be dense, not diagonal. We will not indulge ourselves in the technical details in this regard, save to say that the exchange potential does not appear in our example because our one-particle wave function really is two one-particle states – with identical spatial parts and opposite spins.

The Hartree–Fock wave function of Exercise 4.4.4 is shown in Fig. 4.5(c). Albeit similar, we see that it differs a bit from the one above, which is fixed to be a product of Gaussians. The ability to adjust the shape of the single-particle wave function results in a slightly lower, and thus better, energy estimate. However, neither of the wave functions in Fig. 4.5(a) and (c) are able to account for the fact that the repulsion, Eq. (4.21), reduces the probability of finding both particles at the same spot. As the Hartree–Fock method simply does not pick up this correlation between the particles, the difference between the Hartree–Fock estimate and the actual ground state energy is often referred to as the *correlation energy*.[14] The fact that Egil Hylleraas (Fig. 3.7), was so successful in predicting the helium ground state energy relied on how he was able to explicitly and efficiently accommodate for this correlation in his trial functions.

While the assumption that the ground state could be estimated by a single Slater determinant clearly is a very restrictive one, the Hartree–Fock method often produces decent energy estimates – in particular when the confining potential $V(x)$ is strong compared to the interaction between the particles. When the time-independent Schrödinger equation was able to reveal the mystery behind the periodic table back in the day, the Hartree–Fock method certainly had something to do with that.

[14] It should be mentioned that the precise definition of the term *correlation energy* may differ in different contexts.

One definite advantage of the Hartree–Fock method is that it allows us to tackle the curse of dimensionality by treating a many-particle system as one in which each particle has its own wave function – thus reducing an N-particle problem to N one-particle problems. A dangerous pitfall, on the other hand, is that it may lead us into the temptation of believing that the system actually *is* one in which each particle has its own wave function. This is, however, the case only in the less interesting situation in which the particles do not interact at all. As we discussed in Section 1.3, the true many-body wave function is a lot more complicated.

Single-particle wave functions emerging from Hartree–Fock calculations are frequently used as a starting point for more sophisticated methods, including various flavours of the *configuration interaction* approach mentioned above and other similar techniques such as the *coupled cluster* method.

5 Quantum Physics with Explicit Time Dependence

To some extent you may say that the dynamics we played around with in Chapter 2 is rather trivial; since the Hamiltonian was not actually time dependent, the evolution of the system was directly provided by Eq. (2.19) or, equivalently, Eq. (3.23). When our Hamiltonian actually carries a time dependence, the time evolution must be determined by more sophisticated means. As we will see, it does not have to be exceedingly difficult, though. It may often be achieved as in Eq. (2.27) with slight modifications.

First, however, we will try to keep things simple by playing around with a system with very few degrees of freedom: the spin projection of spin-1/2 particles. Later, things will become a bit more complicated – and interesting – when we address systems for which full spatial dependence comes into play, in addition to explicit time dependence in the Hamiltonian.

There are several situations in which a quantum system is exposed to time-dependent interactions. As we saw in Eq. (2.5), it may be introduced into the Schrödinger equation by exposing the quantum system to an electromagnetic field, such as a laser field. Laser[1] light differs from the electromagnetic radiation emerging from the Sun, for instance, as it consists of one wavelength only and is *coherent* – the light beam as a whole has a well-defined phase at each time and place.[2]

Explicit time dependence also emerges in a quantum system when an atom collides with another particle with a fixed trajectory. Or when the voltages across nano-devices, such as *quantum dots*, are changed in time (see Fig. 5.1).

We will, for the most part, concentrate on the first case – quantum systems exposed to external electromagnetic fields.

5.1 Dynamics with Spin-1/2 Particles

As mentioned, when a particle is confined in space, for instance as illustrated in Fig. 4.4, it makes sense to consider only internal degrees of freedom – such as its spin, given by the *spinor* χ – and disregard the spatial part of the full wave function, see Eq. (4.6), since it does not change. With the spin projection being the only dynamical variable and an external magnetic field $\mathbf{B} = [B_x, B_y, B_z]$ present, the full Hamiltonian is simply the

[1] This acronym stands for *light amplification by stimulated emission of radiation.*

[2] Actually, this quality is crucial for interference patterns such as the one seen in Fig. 1.1 to emerge.

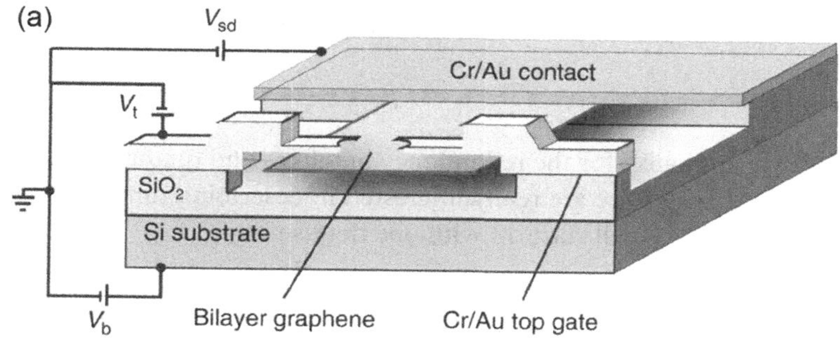

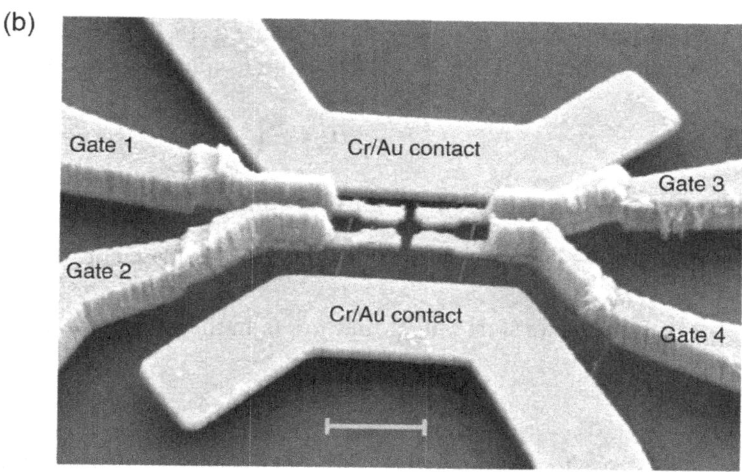

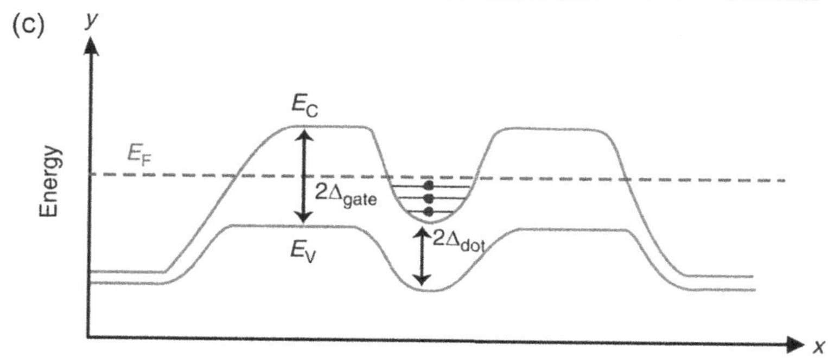

Figure 5.1 By tailoring semiconductor structures and imposing gates with tunable voltages, individual electrons may be trapped in so-called *quantum dots*. Such structures are sometimes referred to as *artificial atoms*; just as in atoms, the energy of the confined electrons is quantized, as indicated in the potential well in (c). Contrary to atoms, however, the energy structure of quantum dots can be manipulated and tuned. The dots are also significantly larger than atoms; the bar in (b) indicates 1 μm (10^{-6} m). The figure is taken from Ref. [3], with permission from Nature Customer Service Centre GmbH.

interaction between spin and the magnetic field. This interaction is given in Eq. (4.15), which we repeat here for convenience:

$$H = -g\frac{q}{2m}\mathbf{B} \cdot \hat{\mathbf{s}} = -g\frac{q\hbar}{4m}\mathbf{B} \cdot \boldsymbol{\sigma} = -\hat{\boldsymbol{\mu}} \cdot \mathbf{B}. \tag{5.1}$$

Here we have introduced the magnetic moment operator:

$$\hat{\mu} = g\frac{q}{2m}\,\hat{\mathbf{s}} = g\frac{q\hbar}{4m}\,\boldsymbol{\sigma}. \tag{5.2}$$

We apologize for the redundancy in names and quantities in these equations.

Although we are really interested in describing time-dependent interactions in this chapter, we will start off with one that is not.

5.1.1 Exercise: Dynamics with a Constant Magnetic Field

Suppose now that the magnetic field in Eq. (5.1) is constant. In that case, with the spinor written as in Eq. (4.13), the Hamiltonian is a 2×2 matrix,[3]

$$H = \frac{1}{2}\begin{pmatrix} -\epsilon & W \\ W^* & \epsilon \end{pmatrix}, \tag{5.3}$$

where the elements ϵ and W are also constants.

(a) How are these constants related to the magnetic field and the parameters in Eq. (5.1)?

Suppose we start out, at time $t = 0$, in the spin-up state,

$$\chi(t=0) = \chi_\uparrow = \begin{pmatrix} 1 \\ 0 \end{pmatrix}.$$

We are to determine the probability of measuring spin up at a later time t:

$$|\langle\chi_\uparrow|\chi(t)\rangle|^2 = \left|\begin{pmatrix} 1 & 0 \end{pmatrix}\begin{pmatrix} a(t) \\ b(t) \end{pmatrix}\right|^2 = |a(t)|^2.$$

For simplicity, let's first assume that the magnetic field points along the x-axis, in which case the Hamiltonian becomes proportional to σ_x.

(b) Under these conditions, explain why W must be real and show, by analytical means, that

$$|a(t)|^2 = \cos^2\left(\frac{Wt}{2\hbar}\right) = \frac{1}{2}\left[1 + \cos\left(Wt/\hbar\right)\right]. \tag{5.4}$$

You may determine $a(t)$ from Eq. (2.19). The exponentiation may be achieved either via diagonalization, as in Eq. (2.23), or by calculating the series, Eq. (2.21), directly. The latter is likely to be more fun; if you make use of what you found in Exercise 4.3.1(a), you will find that the Taylor series for the exponential with our Hamiltonian may be written as a sum of two other Taylor series which may be familiar.

[3] This form is really fixed by Hermicity. Hermicity does not force the diagonal elements to be the negative of each other. However, if they were not, this would only amount to a trivial phase factor – one that we may safely ignore.

In a more general situation in which the magnetic field **B** is allowed to point in *any* direction, the Hamiltonian may be written $H = \mathbf{v} \cdot \boldsymbol{\sigma}$. In this case the matrix exponential may still be done analytically by means rather similar to the above:[4]

$$\exp\left[-\mathrm{i}\mathbf{v}\cdot\boldsymbol{\sigma}\, t/\hbar\right] = \cos\left(vt/\hbar\right)\, I_2 - \mathrm{i}\frac{1}{v}\sin\left(vt/\hbar\right)\,(\mathbf{v}\cdot\boldsymbol{\sigma}), \tag{5.5}$$

where $v = |\mathbf{v}|$.

(c) Verify that the Hamiltonian of Eq. (5.3) may be written

$$H = \frac{1}{2}\left(-\varepsilon\sigma_z + \mathrm{Re}\, W\,\sigma_x - \mathrm{Im}\, W\,\sigma_y\right) \tag{5.6}$$

Use this and Eq. (5.5) to show that for a general situation in which ϵ is not necessarily zero and W not necessarily real,

$$|a(t)|^2 = \frac{1}{\epsilon^2 + |W|^2}\left[\epsilon^2 + |W|^2\cos^2\left(\frac{\sqrt{\epsilon^2+|W|^2}}{2\hbar}\, t\right)\right]. \tag{5.7}$$

(d) According to Eq. (5.7), what does it take to achieve a complete spin flip, $a(t) = 0$? What restrictions must we impose on the parameters ϵ, W and the duration t of the interaction to make this happen?

As mentioned, this exercise was not entirely consistent with the name of the chapter. So, for the next one, we'll make sure to include time-dependent terms in our Hamiltonian.

5.1.2 Exercise: Dynamics with an Oscillating Magnetic Field

In Exercise 5.1.1 we saw that with a constant magnetic field the spin projection oscillates following a cosine. The dynamics is richer if we expose the particle to a time-varying magnetic field. With a static field B_z in the negative z-direction and an oscillating field $B(t)$ in the xy-plane, the Hamiltonian may be written[5]

$$H = \begin{pmatrix} -\epsilon/2 & \hbar\Omega\sin(\omega t) \\ \hbar\Omega^*\sin(\omega t) & \epsilon/2 \end{pmatrix}. \tag{5.8}$$

(a) With $\hbar = 1$, set the parameters to $\epsilon = 1$, $\Omega = 0.2$ and $\omega = 1.1$ and, again, take your initial state to be $\chi_\uparrow$. In other words, set $a(t = 0)$ to 1 and $b(t = 0)$ to 0. Then, using an ODE solver you are familiar with, solve the Schrödinger equation

$$\mathrm{i}\hbar\frac{\mathrm{d}}{\mathrm{d}t}\chi = H\chi, \tag{5.9}$$

with χ given by Eq. (4.13) and H provided in Eq. (5.8).

[4] You are more than welcome to prove this. In doing so, it may be useful to note that for a general vector $\mathbf{v}$, $\left(\mathbf{v}\cdot[\sigma_x,\sigma_y,\sigma_z]\right)^2 = v^2$, so that $(\mathbf{v}\cdot\boldsymbol{\sigma})^{2n} = v^{2n}$ and $(\mathbf{v}\cdot\boldsymbol{\sigma})^{2n+1} = v^{2n+1}\,\mathbf{v}/v\cdot\boldsymbol{\sigma}$.

[5] The factor $\hbar$ in the off-diagonal elements may appear somewhat odd at present. It is convenient to introduce the factor here for reasons that, hopefully, will become apparent later on.

(b) Why can't we do as we did in Exercise 5.1.1 and solve the Schrödinger equation by means of Eq. (2.19)?

(c) Plot the spin-up probability $|a(t)|^2$ as a function of time.

(d) As a check of your implementation, see that you reproduce Eq. (5.7) numerically when you replace $\sin(\omega t)$ by $1/2$ in Eq. (5.8). (With W in Eq. (5.3) equal to $\hbar\Omega$.)

(e) Shifting back to the time-dependent Hamiltonian, play around with the parameters a bit. Are you able to obtain a more or less complete spin flip in this case even when $\epsilon \neq 0$?

Although this exercise addresses a rather simple system, it is frequently used to describe more advanced ones as well. The Hamiltonian of Eq. (5.8) – and simple extensions of it – is for instance often used to model atoms and ions exposed to laser fields. This is admissible if the dynamics somehow involves transitions between two atomic/ionic states only.

Actually, such simulations tend to be made even simpler by substituting the time-dependent Hamiltonian (5.8) with the static one in Eq. (5.3). If you find the notion of describing the interaction with a time-dependent field by means of a time-independent Hamiltonian rather odd, we can only agree. There is a trick, however, which does allow you to do so under certain conditions. The trick is called *the rotating wave approximation*.

5.1.3 Exercise: The Rotating Wave Approximation

Our starting point is exactly the same as in Exercise 5.1.2 – the time-dependent Schrödinger equation for a spin-1/2 particle in an external field, whose Hamiltonian is given by Eq. (5.8).

(a) First, we define an alternative representation of the spin state by imposing a unitary transformation, $\chi' = U\chi$, where

$$U = \begin{pmatrix} e^{-i\epsilon t/2\hbar} & 0 \\ 0 & e^{i\epsilon t/2\hbar} \end{pmatrix}. \tag{5.10}$$

Verify that U is, in fact, unitary; $U^\dagger = U^{-1}$.

(b) The state χ' now follows a Schrödinger equation with the modified Hamiltonian

$$H' = \begin{pmatrix} 0 & \hbar\Omega \sin(\omega t) e^{-i\epsilon t/\hbar} \\ \hbar\Omega^* \sin(\omega t) e^{i\epsilon t/\hbar} & 0 \end{pmatrix}. \tag{5.11}$$

Verify this by setting up the time-dependent Schrödinger equation, Eq. (5.9), with the Hamiltonian of Eq. (5.8) and replacing χ with $U^\dagger \chi'$.

Hint: Remember that also the left hand side of the Schrödinger equation will contribute to H'.

(c) If we write $\sin(\omega t)$ as $(e^{\mathrm{i}\omega t} - e^{-\mathrm{i}\omega t})/2\mathrm{i}$, the non-zero elements of the Hamiltonian in Eq. (5.11) will acquire two terms. If $\hbar\omega \approx \epsilon$, one of these will oscillate slowly and the other one will oscillate quite rapidly in comparison. Due to this, the latter may often be neglected.[6]

Show that this approximation leads to the effective Hamiltonian

$$H' \sim \frac{\hbar}{2}\begin{pmatrix} 0 & -\mathrm{i}\Omega e^{-\mathrm{i}\delta t} \\ \mathrm{i}\Omega^* e^{\mathrm{i}\delta t} & 0 \end{pmatrix}, \tag{5.12}$$

where the parameter

$$\delta = \epsilon/\hbar - \omega. \tag{5.13}$$

(d) Along the same lines as in (b), impose the unitary transformation

$$V = \begin{pmatrix} e^{+\mathrm{i}\delta t/2} & 0 \\ 0 & e^{-\mathrm{i}\delta t/2} \end{pmatrix} \tag{5.14}$$

on χ' to arrive at yet another way of representing the spin state, $\chi'' = V\chi'$.

Show that this leads to the time-independent Hamiltonian

$$H_{\mathrm{RWA}} = \frac{\hbar}{2}\begin{pmatrix} -\delta & -i\Omega \\ i\Omega^* & \delta \end{pmatrix}. \tag{5.15}$$

This is, in fact, the same as the Hamiltonian of Eq. (5.3) – with the correspondence $\delta \sim \epsilon/\hbar$ and $\Omega \sim \mathrm{i}W/\hbar$.

(e) We have now imposed two transformations on χ; the Hamiltonian of Eq. (5.15) relates to the state

$$\chi'' = V\chi' = VU\chi. \tag{5.16}$$

Verify that the components of χ'' and χ satisfy $|a''(t)|^2 = |a(t)|$ and $|b''(t)|^2 = |b(t)|$. This means that we do not need to transform back to the original χ-representation in order to determine the spin-up and spin-down probabilities.

(f) Check whether our analytical solution for the constant Hamiltonian, Eq. (5.7), actually agrees, more or less, with your numerical solution from Exercise 5.1.2. How large must δ be before the rotating wave approximation breaks down – before our analytical, approximate solution starts deviating significantly from the correct one? Also, try to figure out, by numerical tests, how the validity of the rotating wave approximation is affected by the magnitude of Ω.

As mentioned, a spin-1/2 particle in a constant magnetic field is not the only system for which the Hamiltonian of Eq. (5.3) is encountered; it has several applications. For an atom or an ion driven between two states by a laser field, with very little population of any other states, Eq. (5.8) is a decent approximation. If the angular frequency ω is

[6] Simply put, its contribution integrates to zero – pretty much in the same way as $\int_a^b f(x)\sin(kx)\,dx \approx 0$ if $f(x)$ changes little during one period of the sine function and the integration interval extends over several periods.

such that it is resonant or near-resonant with the transition between the two states, $\hbar\omega \approx \epsilon$, and the coupling strength $|\Omega|$ is comparatively low, we can approximate such a system by the constant Hamiltonian of Eq. (5.3), or, equivalently, Eq. (5.15), via the rotating wave approximation. In fact, several works exploit this in order to study how matter may be manipulated by electromagnetic radiation.

In such a context, $|\Omega|$ or $|W|/\hbar$ is frequently is referred to as the *Rabi frequency*, named after Isodor Isaac Rabi, who won the 1944 Nobel Prize in Physics for the discovery of *nuclear magnetic resonance*, which we will address in Chapter 6. The Rabi frequency is related to the so-called dipole transition element:

$$\Omega = -\frac{q}{\hbar} \langle \psi_0 | \mathbf{E}_0 \cdot \mathbf{r} | \psi_1 \rangle, \tag{5.17}$$

where q is the charge of the particle, $\psi_{0,1}$ are the wave functions corresponding to the two atomic states involved, and $\mathbf{E}_0$ gives the amplitude and direction of the oscillating electric field set up by the laser. The δ-parameter in Eq. (5.13), which is called the *detuning*, is the difference between the energy gap between the two states, ϵ, and the energy of one photon, $\hbar\omega$, divided by $\hbar$. The quantity

$$\Omega_\mathrm{G} = \sqrt{\delta^2 + |\Omega|^2} \tag{5.18}$$

is called the *generalized Rabi frequency*. This is the angular frequency that appears in Eq. (5.7). If we rewrite this equation in terms of detuning and Rabi frequencies, it reads

$$|a(t)|^2 = \frac{1}{|\Omega_\mathrm{G}|^2} \left[\delta^2 + |\Omega|^2 \cos^2 \left(\Omega_\mathrm{G}\, t \right) \right]. \tag{5.19}$$

We will now add one more particle. You have been warned about the curse of dimensionality earlier. However, the spin projection of spin-1/2 particles resides in a two-dimensional space. In going from one such particle to two, the dimensionality increases from 2^1 to 2^2, which really isn't much of a curse.

5.1.4 Exercise: Spin Dynamics with Two Spin-1/2 Particles

In the model we study in this exercise, each of the two particles is exposed to the same magnetic fields as in Exercise 5.1.2. We also introduce an interaction between the particles; the spin of each particle has a certain influence on the other.[7] The interaction energy differs by u' depending on whether the spins are aligned or anti-aligned. In a physical implementation, the magnitude of u' would depend on the distance between the two fixed spin-1/2 particles – and on the nature of the material in which the system is embedded.

[7] This interaction comes about because each particle with spin sets up a small magnetic field which affects the spin of the other.

We write the total two-particle spin state χ_2 as in Eq. (4.10) – a linear combination of the product states in Eqs. (4.9). With reference to this basis, we may write χ_2 as a vector in $\mathbb{C}^4$:

$$\chi_2 = a\, \chi_\uparrow^{(1)} \chi_\uparrow^{(2)} + b\, \chi_\uparrow^{(1)} \chi_\downarrow^{(2)} + c\, \chi_\downarrow^{(1)} \chi_\uparrow^{(2)} + d\, \chi_\downarrow^{(1)} \chi_\downarrow^{(2)} \rightarrow \begin{pmatrix} a(t) \\ b(t) \\ c(t) \\ d(t) \end{pmatrix} \tag{5.20}$$

and our Hamiltonian becomes a 4×4 matrix:

$$H = h^{(1)} + h^{(2)} + u'\, \mathbf{s}^{(1)} \cdot \mathbf{s}^{(2)} \tag{5.21a}$$

$$= \begin{pmatrix} -\epsilon + u & A & A & 0 \\ A^* & -u & 2u & A \\ A^* & 2u & -u & A \\ 0 & A^* & A^* & +\epsilon + u \end{pmatrix}, \tag{5.21b}$$

where $h^{(k)}$ is the single-particle Hamiltonian of Eq. (5.8) for particle k. We have introduced the quantities $A(t) \equiv \Omega \sin(\omega t)$ and $u \equiv u'\hbar^2/4$ for convenience. Actually, while the matrix in Eq. (5.21b) is accurate, we may be accused of being somewhat sloppy for writing the full two-particle Hamiltonian as in Eq. (5.21a). The single-particle parts should, as in Exercise 4.4.3, really be the tensor/Kronecker product with the identity operator for the other particle; $h^{(1)}$ and $h^{(2)}$ really mean $h \otimes I_2$ and $I_2 \otimes h$, respectively. The spin–spin interaction is also given by tensor products: $\mathbf{s}^{(1)} \cdot \mathbf{s}^{(2)} = (\hbar/2)^2\, (\sigma_x \otimes \sigma_x + \sigma_y \otimes \sigma_y + \sigma_z \otimes \sigma_z)$.

(a) Take your initial state to be the one in which both spins point upwards and solve the Schrödinger equation using your ODE solver of preference. You can use the same values for ϵ, Ω and ω as in Exercise 5.1.2 and set $u = 0.025$. As functions of time, plot the probability for both spins to point upwards, $|a(t)|^2$, and the probability for finding the system in state $\chi_\uparrow^{(1)} \chi_\downarrow^{(2)}$, $|b(t)|^2$, where a and b refer to Eq. (5.20).

(b) If our particles are identical, is the initial condition in (a) bona fide? Is this an admissible state considering the exchange symmetry? For a general situation in which our initial spin state has a well-defined exchange symmetry, does the Schrödinger equation with the Hamiltonian of Eq. (5.21) maintain this symmetry? Will the spin state necessarily have the same symmetry at a later time t?

Try to answer this question both by checking your numerical solution at different times and, more theoretically, by studying the form of the Hamiltonian.

(c) Rerun your calculation using the last triplet state in Eqs. (4.11a) and the singlet state, Eq. (4.11b), as initial conditions instead of $\chi_2(t = 0) = \chi_\uparrow^{(1)} \chi_\uparrow^{(2)}$. One of these runs is rather boring, isn't it? Why is that?

Hint: Perhaps the lessons learned in the last parts of Exercise 2.6.2 may shed some light on the matter?

(d) Now we lift the restriction that the particles are to be identical and take our initial state to be $\chi_\uparrow^{(1)}\chi_\downarrow^{(2)}$. Rerun your calculation with various values for the spin–spin interaction strength u and see how this strength affects the spin dynamics.

How do the spin dynamics depend on u if you return to an exchange symmetric initial state?

In Chapter 6 we will address quantum technologies. One prominent example of these is the *quantum computer*. Instead of 'traditional' bits, zeros and ones, it uses *quantum bits* as information units. A quantum bit, a *qubit*, is simply a dynamical quantum system consisting of two states. Thus, the spin projection of a spin-1/2 particle may very well constitute a qubit. And what we did in Exercises 5.1.2 and 5.1.4 could be considered a simulation of something going on in a quantum computer.

Now we will turn to more complex dynamics – dynamics involving spatial wave functions.

5.2 Propagation

The evolution of the wave function from a time t to a later time $t + \Delta t$ may be written as the action of an operator:

$$\Psi(t + \Delta t) = \hat{U}(t, t + \Delta t)\Psi(t). \tag{5.22}$$

This operator, $\hat{U}$, is referred to as a *propagator*. We are already familiar with what the propagator looks like with a time-independent Hamiltonian, see Eqs. (2.19) and (2.27). We have also learned, however, that when $\hat{H}$ is time dependent, these equations no longer hold. As mentioned, Eq. (2.27) may still be useful, though.

5.2.1 Exercise: Magnus Propagators of First and Second Order

(a) Show that the error in using Eq. (2.27) for the time evolution is proportional to Δt^2 at each time step. To this end, differentiate Eq. (2.1) in order to express the first three terms in a Taylor expansion of $\Psi(t + \Delta t)$ and compare them to the first three terms in an expansion of Eq. (2.27). See Eq. (2.20).

(b) Show that if you replace $\hat{H}(t)$ in Eq (2.27) by $\overline{H} = 1/\Delta t \int_t^{t+\Delta t} \hat{H}(t')\,\mathrm{d}t'$, the error is proportional to Δt^3.

Hint: Use Taylor expansion or the trapezoidal rule to approximate the integral.

Let's summarize:

$$\Psi(t + \Delta t) = e^{-\mathrm{i}\,H(t)\Delta t/\hbar}\Psi(t) + \mathcal{O}(\Delta t^2), \tag{5.23a}$$

$$\Psi(t + \Delta t) = e^{-\mathrm{i}\,\overline{H}(t)\Delta t/\hbar}\Psi(t) + \mathcal{O}(\Delta t^3), \tag{5.23b}$$

$$\text{where} \quad \overline{H}(t) \equiv \frac{1}{\Delta t}\int_t^{t+\Delta t} \hat{H}(t')\,\mathrm{d}t'.$$

So we get a significant improvement in accuracy if we replace the Hamiltonian with its time average over the interval in question. Since the error, the *local* one that is, is of third order, the time-averaged Hamiltonian $\overline{H}$ may safely be approximated by the trapezoidal or midpoint rule:

$$\overline{H} \approx \frac{1}{2}\left(\hat{H}(t) + \hat{H}(t + \Delta t)\right) \approx \hat{H}(t + \Delta t/2). \tag{5.24}$$

Equations (5.23) are examples of *Magnus propagators*, specifically Magnus propagators of first and second order, respectively. They are named after the German–American mathematician Wilhelm Magnus. Approximating the exact propagator, Eq. (5.22), as an exponential with a, possibly, time-averaged Hamiltonian in this way is convenient in several respects. One advantage of approximating the propagator in this way is its norm-conserving property; although we introduce errors when using Eqs. (5.23) with finite Δt, the *norm* of the wave function is always preserved to machine accuracy. Another advantage is the fact that they provide *explicit* schemes for resolving the time evolution – as opposed to implicit schemes in which you have to solve an equation to get from one time to the next.

There is a drawback with propagators of the form (5.23), however. Since the Hamiltonian now changes in time, we cannot exponentiate it once and then use it repeatedly as we did in Chapter 2. We must perform a full exponentiation at each time step. This is quite costly, we really don't want to do that. Luckily, there are ways around it.

5.2.2 Exercise: Split Operators

(a) While the relation $e^{a+b} = e^a e^b$ holds for any numbers a and b, why is this generally not the case with operators or matrices; $e^{\hat{A}+\hat{B}} \neq e^{\hat{A}} e^{\hat{B}}$?

(b) Use Taylor expansion to prove that

$$e^{(\hat{A}+\hat{B})\,\Delta t} = e^{\hat{A}\,\Delta t/2} e^{\hat{B}\,\Delta t} e^{\hat{A}\,\Delta t/2} + \mathcal{O}(\Delta t^3). \tag{5.25}$$

When we want to approximate a Magnus propagator, Eq. (5.25) may be very useful indeed as it allows us to treat the various parts of the exponentiation of the Hamiltonian independently – and yet maintain a reasonable degree of precision.

In many cases, the Hamiltonian may be split into a time-dependent, part and a time-independent part:

$$\hat{H} = \hat{H}_0 + \hat{H}'(t). \tag{5.26}$$

This is, for instance, the situation for a quantum particle exposed to a strong laser field. The time-independent part, $\hat{H}_0$, is just $\hat{T} + V$ as before while the laser interaction, in one dimension, may be written

$$\hat{H}'(t) = -qE(t)x. \tag{5.27}$$

You may recognize this interaction from Eq. (5.17); q is the charge of the particle and $E(t)$ is the time-dependent strength of the electric field of the laser.[8]

With the Hamiltonian in Eq. (5.26), with $\hat{H}'$ time averaged, $\hat{A} = -\mathrm{i}\hat{H}_0/\hbar$ and $\hat{B} = -\mathrm{i}\overline{H'}/\hbar$, we may estimate the wave function at the next time step as

$$\Psi(t + \Delta t) = e^{-\mathrm{i}\hat{H}_0\,\Delta t/2\hbar} e^{-\mathrm{i}\overline{H'}\,\Delta t/\hbar} e^{-\mathrm{i}\hat{H}_0\,\Delta t/2\hbar}\,\Psi(t) + \mathcal{O}(\Delta t^3). \tag{5.28}$$

This means that the time-independent $\hat{H}_0$-part may be exponentiated initially – once and for all – and used repeatedly. The time-dependent part, $\overline{H'}$, on the other hand, must be exponentiated anew for each time step. This is not necessarily bad news, however, since quite often this is easily implemented. For instance, the interaction in Eq. (5.27) is diagonal in a position representation. Thus, exponentiating it is trivial – you just exponentiate the diagonal elements directly.

It should be mentioned that both Magnus propagators and split-operator techniques of higher order than three in (local) error are well established. Third order will do for our purposes, however.

5.2.3 Exercise: Photoionization

The time is ripe to actually solve the time-dependent Schrödinger equation for a spatial wave function again. We will do so for a one-dimensional model for an atom exposed to a short, intense laser pulse:

$$E(t) = \begin{cases} E_0 \sin^2\left(\frac{\pi}{T}t\right)\sin(\omega t), & 0 \leq t \leq T, \\ 0, & \text{otherwise}. \end{cases} \tag{5.29}$$

We set the charge q in Eq. (5.27) to -1 in our convenient units. Here E_0 is the maximum field strength, T is the duration of the pulse and ω is the central angular frequency of the laser. Although we do not really consider photons within this approximation, it still makes sense to relate the angular frequency ω to the photon energy via $E_\gamma = \hbar\omega = hf$, see Eq. (1.4).

You can choose the field parameters $\omega = 1$, $E_0 = 1$ and $T = 10\,t_{\text{cycl}}$, where $t_{\text{cycl}} = 2\pi/\omega$ is a so-called *optical cycle* (see Fig. 5.2). Physically, this would correspond to a very intense pulse in the extreme ultraviolet region.

Also, apply a static potential of the form of Eq. (2.30) with $V_0 = -1$, $s = 5$ and $w = 5$. As this exercise involves a number of parameters – both numerical and physical ones – we summarize them in the table below. The set of parameters, which includes a few more than the ones listed above, is given in our usual units – apart from the pulse duration T, which is given in terms of optical cycles.

E_0	ω	T	ΔT	V_0	w	s	Δt	L	$n+1$
0.5	1	$10\,t_{\text{cycl}}$	25	-1	5	5	0.1	400	512

[8] Here we have neglected the fact that the field **E** would also be position dependent not just time dependent. Often this is admissible since the extension of the system in question, such as an atom or a molecule, is much smaller than the wavelength of the electric field. This approximation is referred to as the *dipole approximation*.

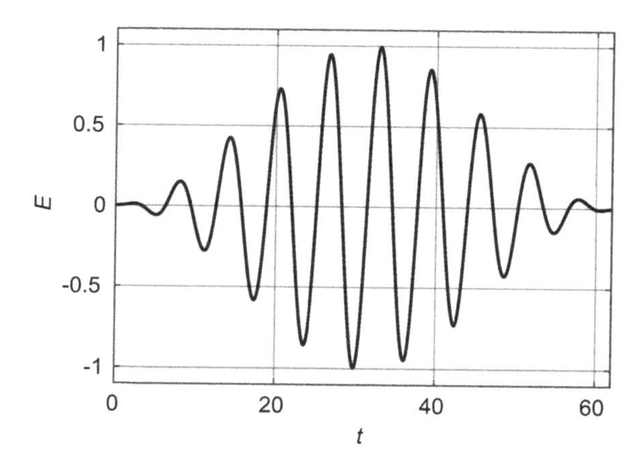

Figure 5.2 In Exercise 5.2.3 we expose our model atom to a time-dependent electric field with this shape. Both the amplitude E_0 and the angular frequency ω are 1 in our units. The pulse duration T corresponds to 10 so-called *optical cycles*, $T = 10 \cdot 2\pi/\omega$.

We will adjust some of the numerical parameters in the following. The fact that we suggest starting out with the number of grid points $N = n + 1 = 2^9$ reveals our preference when it comes to approximating the kinetic energy here.[9]

Just as we ignored spatial degrees of freedom above, we will ignore spin in this context. This is admissible if we start out in a product state, Eq. (4.6), since our Hamiltonian does not depend on spin.

(a) By using either propagation in imaginary time or direct diagonalization, find the ground state of the time-independent system – the eigenfunction of H_0 with the lowest energy. Let this be your initial state, $\Psi(x;t = 0)$. Check that the ground state energy is well converged in terms of the number of grid points – that it does not change much if you increase n.

(b) Implement the propagator in Eq. (5.28) and use it to simulate your system for the entire duration of the laser pulse – and for some time ΔT beyond. As in Chapter 2, plot $|\Psi(x;t)|^2$ as you go along. Since most of the wave packet will remain bound – only a limited fraction is liberated by the laser pulse – you may want to adjust the upper limit of your y-axis so that you see the outgoing waves clearly. An upper limit of 10^{-2} units should be adequate here. Check that the grid size $L = 400$ length units is large enough to avoid the artefacts we saw in Exercise 2.3.2.

For the numerical time step, you can set $\Delta t = 0.1$ initially. Note the difference between this time parameter and ΔT, which is the duration of the simulation after the pulse, Eq. (5.29), has passed. The latter can be fixed at 25 time units.

(c) Our initial grid is a rather crude one. Increase the number of grid points N and rerun your simulation to check for convergence. Also, rerun your simulation with

[9] In plain words, we suggest using the FFT version.

smaller time step Δt to check whether it has been set small enough. Why was this not necessary when we did similar things in Chapter 2, by the way?

(d) How much of your wave packet has actually left the region of the confining potential? Or, in other words, what is the probability for our model atom to be *ionized* by the laser pulse? This probability can be found by means similar to the way we determined transmission and reflection probabilities in Exercise 2.4.1. Just make sure that no bound part of the wave function contributes to our ionization probability estimate.

What we just did is a rather direct approach to determining the ionization probability. The term *ionization* refers to the fact that, after tearing loose an electron from an atom, the remainder of the atom is charged, it is an *ion*.

In the following we will try to arrive at the same probability in a slightly different way. We will also see how we may extract additional information about the unbound part of the wave packet.

5.3 Spectral Methods

Spectral methods represent a rather general approach to solving partial differential equations. Arguably, they are particularly well suited to quantum physics. One reason for this is the fact that the Schrödinger equation is *linear*; the Hamiltonian is a linear operator which transforms the wave function to the same space. Consequently the wave function may be written as a linear combination of a set of basis states at all times:

$$\Psi(t) = \sum_n a_n(t)\,\varphi_n. \tag{5.30}$$

Now, if we have a basis, $\{\varphi_i\}$, for the vector/function space in which our wave function resides, and an inner product, such as the one defined in Eq. (1.24), we may use this to turn the Schrödinger equation, which really is a partial differential equation, a *PDE*, into an ordinary differential equation, an *ODE*. This, in turn, will allow us to make use of ODE machinery we already know – such as *Runge–Kutta* methods, for instance.

We start out by expressing our wave function as in Eq. (5.30) and inserting it into the Schrödinger equation, Eq. (2.1). We assume our basis to be discrete/countable and finite. And, for now, we assume the basis functions $\varphi_n(x)$ to be time independent and keep all time dependence in the coefficients $a_n(t)$. Next, we impose inner products on each and every one of the basis states φ_k from the left at both sides of the equation. As we will see in the next exercise, this produces a set of coupled differential equations in time – and time only.

5.3.1 Exercise: The Matrix Version

(a) Explain how the procedure described above leads to the following set of equations:

$$i\hbar \sum_n \langle \varphi_k | \varphi_n \rangle \, a_n'(t) = \sum_n \langle \varphi_k | \hat{H} | \varphi_n \rangle \, a_n(t). \tag{5.31}$$

(b) Explain how this may be written in matrix form as

$$i\hbar S \frac{\mathrm{d}}{\mathrm{d}t} \mathbf{a} = H \mathbf{a}. \tag{5.32}$$

What do the elements of the matrices S and H look like?

(c) Why does the above equation simplify when the basis we use is orthonormal? See Eq. (3.19).

In a practical implementation, it is not a prerequisite that the basis states used to represent our wave function in Eq. (5.30) are orthonormal. Non-orthogonal basis functions, such as Gaussians or so-called *b-splines*, are frequently applied. However, if the basis functions do fulfil Eq. (3.19), the matrix on the left hand side of Eq. (5.32), which is called the *overlap matrix*, is just the identity matrix and the left hand side of Eq. (5.32) becomes simply $i\hbar \mathbf{a}'(t)$. We exploit this advantage if we take our basis to be the eigenstates of the time-independent part of the Hamiltonian $\hat{H}$, that is, the $\hat{H}_0$ of Eq. (5.26). As we have stated earlier, particularly explicitly in Section 3.3, the Hamiltonian is Hermitian, and thus we can always construct an orthonormal basis consisting of its eigenstates. Another advantage with this choice is that $\hat{H}_0$ becomes a diagonal matrix in this representation.

And there is even a third advantage. Such a choice of basis often makes analysing the wave function rather straightforward. After the interaction is over, when $t > T$, our Hamiltonian is, again, time independent and our wave function evolves according to Eq. (3.23):

$$\Psi(t > T) = \sum_n a_n(t = T) e^{-i\varepsilon_n (t-T)/\hbar} \, \psi_n. \tag{5.33}$$

With ψ_n being the nth eigenstate of $\hat{H}_0$, see Eq. (3.18), an energy measurement would provide the result ε_n with probability

$$\begin{aligned} P(\varepsilon_n) &= |\langle \psi_n | \Psi(t > T) \rangle|^2 = \left| \sum_k a(t = T) e^{-i\varepsilon_n (t-T)/\hbar} \langle \psi_n | \psi_k \rangle \right|^2 \\ &= \left| a(t = T) e^{-i\varepsilon_n (t-T)/\hbar} \right|^2 = |a_n(T)|^2, \end{aligned} \tag{5.34}$$

see Eq. (2.41).

With the normalized eigenstates of H_0 as our basis, the remaining equation to solve is

$$i\hbar \frac{\mathrm{d}}{\mathrm{d}t} \mathbf{a} = \left(\mathrm{Diag}(\varepsilon_0, \varepsilon_1, \ldots) + H' \right) \mathbf{a} \tag{5.35}$$

with the time-dependent matrix elements

$$H'_{k,n} = \langle \psi_k | \hat{H}' | \psi_n \rangle. \tag{5.36}$$

Equation 5.35 can be solved by any ODE solver you know. But that is not to say that all of them are equally well suited.

5.3.2 Exercise: Dynamics Using a Spectral Basis

Redo Exercise 5.2.3 using a spectral basis consisting of the eigenstates of $\hat{H}_0$. You are first to construct a numerical representation of $\hat{H}_0$, diagonalize it to find the eigenstates/eigenvectors, ensure proper normalization, and use these as your basis.

Next, construct a matrix representation of $\hat{H}'$; this means that all matrix elements of type $\langle \psi_k | x | \psi_n \rangle$, which are to be multiplied by $-qE(t)$, must be calculated. This can be done quite efficiently by means of matrix products between matrices involving the eigenstates of H_0 and a diagonal matrix with the x-values in your grid. Do remember that with the eigenstates of H_0 being normalized according to Eq. (2.10), you will need to multiply this matrix product by the spatial increment h. As for the grid, the extension and the number of grid points you found to be adequate in Exercise 5.2.3 are still adequate.

In resolving the dynamics, use any ODE solver of your own preference. You can, for instance, use some familiar Runge–Kutta method – preferably one with an adaptive time step Δt, one that adjusts Δt at each step in order to ensure reasonable precision. If, on the other hand, you do this from scratch with a fixed time step, chances are that your time step needs to be very small. The backward–forward Euler method, or the *Crank–Nicolson method*, may be more apt for this task:

$$\left[\hat{I} + \frac{\mathrm{i}}{\hbar}\hat{H}(t+\Delta t)\frac{\Delta t}{2}\right]\mathbf{a}(t+\Delta t) \approx \left[\hat{I} - \frac{\mathrm{i}}{\hbar}\hat{H}(t)\frac{\Delta t}{2}\right]\mathbf{a}(t). \tag{5.37}$$

Here $\hat{I}$ is the identity operator. Along with split-operator techniques and other approximations to Magnus-type propagators, this is a much applied method in quantum dynamics. Note that this is an implicit scheme; you will need to solve an equation by matrix inversion at each time step to obtain $\mathbf{a}(t + \Delta t)$ explicitly This propagator would also work with our usual grid representation. And also, with proper adjustments, the split-operator scheme would still work within this spectral approach.

This time, do not bother to display the evolution of the wave function, just determine the vector with expansion coefficients at time $t = T$, $\mathbf{a}(T)$.

Now, the probability for your system to remain bound – to remain in an eigenstate with negative energy – after the interaction may, according to Eq. (5.34), be calculated as

$$P_{\text{bound}} = \sum_{\varepsilon_n < 0} |a_n(T)|^2 . \tag{5.38}$$

Calculate this probability – and the corresponding ionization probability,

$$P_{\text{ionized}} = 1 - P_{\text{bound}}. \tag{5.39}$$

Does it agree with what you found in Exercise 5.2.3?

Why is that here, contrary to Exercise 5.2.3, we can stop our time propagation at $t = T$?

What is the probability of remaining in the ground state? And what is the probability of exciting the 'atom' – to promote it from the ground state to another bound state? Finally, if you remove some of the states with the highest eigenenergies from our spectral basis, how does this affect the result?

We have now, in two equivalent ways, calculated the probability of liberating our quantum particle from its confining potential. However, we may also want to know how this liberated wave packet is distributed in energy or *momentum*. This is typically something you would want to measure in an experiment; the energy distribution of the liberated electrons carries a lot of information about the processes that lead to ionization and about the structure of the atom or molecule in question. Since we have already expressed our wave function as a linear combination of eigenstates of $\hat{H}_0$, it would seem like the natural choice to use the expansion coefficients corresponding to positive energies, $|a_n(T)|^2$ with $\varepsilon_n > 0$, for this purpose. However, it is not that straightforward – for reasons we will get back to shortly.

As an alternative, we can determine how the liberated particle is distributed in energy or momentum by calculating its Fourier transform, which we introduced in Eq. (2.12). The variable of the Fourier-transformed wave function, k, is proportional to the momentum p, $k = p/\hbar$. So, with our usual choice of units, in which $\hbar = 1$, they are in fact the same. Now, in the same way that $|\Psi(x)|^2$ provides the position distribution of the quantum particle, $|\Phi(k)|^2$, with $\Phi(k) = \mathcal{F}\{\Psi\}(k)$, provides its momentum distribution. More precisely,

$$\frac{\mathrm{d}P}{\mathrm{d}p} = \hbar^{-1}|\Phi(p/\hbar)|^2 \tag{5.40}$$

is the probability density for a momentum measurement to yield the result p. This is the Born interpretation, which we first encountered in regard to Eq. (1.9), formulated in terms of momentum instead of position.

In order to determine the momentum distribution for the liberated part of the wave packet exclusively, we must first get rid of the part that is still bound. Moreover, we will be interested in the momentum or the energy the liberated particle has *asymptotically* – away from the confining potential $V(x)$. We don't want the potential to influence the momentum distribution. As we will see, these issues can be resolved by propagating ΔT for a while beyond the duration of the laser interaction – as we did in Exercise 5.2.3.

5.3.3 Exercise: Momentum Distributions

Here we will implement the procedure we just outlined – again using the system of Exercise 5.2.3 as example. When imposing the Fourier transform, using your FFT implementation, you should be aware of two potential issues:

(1) as discussed in Section 2.2, standard FFT implementations typically permute the k-vector in a somewhat odd way, and

(2) although the norm of $\Phi(k)$ in 'k-space' should be the same as the norm of $\Psi(x)$ in 'x-space',

$$\int_{-\infty}^{\infty} |\Phi(k)|^2 \, \mathrm{d}k = \int_{-\infty}^{\infty} |\Psi(x)|^2 \, \mathrm{d}x, \tag{5.41}$$

the numerical implementation you choose to use may not ensure this.

(a) Rerun your implementation from Exercise 5.2.3, and check that your wave function after the laser interaction has a clear distinction between a bound part and an unbound part. Perhaps you would need a larger box size L and longer after-propagation ΔT to ensure this. By inspecting your final wave function, determine an appropriate value of d that separates the bound and the unbound parts; in other words, any part of the wave function for which $|x| > d$ can be interpreted as ionized or liberated. This requires that the simulation is allowed to run long enough for everything unbound to travel beyond $x = \pm d$.

Your d value should be set large enough to avoid the influence of the confining potential. In other words, $V(x) \approx 0$ for all $|x| > d$.

(b) Set $\Psi(x)$ to zero for all $|x| < d$ and Fourier transform the remaining part. Plot the corresponding momentum distribution:

$$\frac{\mathrm{d}P}{\mathrm{d}p} = \hbar^{-1} \left| \mathcal{F}\left\{\Psi(|x| > d)\right\}(k) \right|^2 . \tag{5.42}$$

(c) Next, repeat the simulation but this time with two consecutive laser pulses,

$$E_{\mathrm{Two}} = E(t) + E(t - (T + \tau)), \tag{5.43}$$

where $E(t)$ refers to Eq. (5.29) and τ is the time delay between the pulses. Afterwards, in the same manner as above, plot the momentum distribution of the liberated particle at some time $t > 2T + \tau$. In order to avoid numerical artefacts, you may need to increase your box size further. Such an increase should be accompanied by a corresponding increase in the number of grid points $n + 1$.

This time, does the momentum distribution resemble twice what you got with only one pulse? Should it? Does the distribution depend on the time delay τ? What do we call this phenomenon?

This last exercise touches upon an issue we have shamelessly swept under the rug thus far. In every example that addresses spectral bases, we have exclusively dealt with *countable* and *finite* basis sets. Numerically, this makes sense because we, artificially, impose a finite extension of our numerical grid with some boundary condition combined with a finite number of grid points. Physically, however, it makes less sense. The energy of a free particle is not quantized; eigenenergies of an unbound particle are continuous and unbound, not discrete and finite. To wit, the momentum distributions we just calculated are continuous, we could use them to obtain energy distributions that would be continuous too. Correspondingly, Eq. (5.34) does not really work unless we are dealing with discrete energies – the quantized part of the energy spectrum. For a true energy continuum, a different formulation is called for.

Having said this, it should also be said that our numerical *pseudo*-continuum, which is artificially discretized due to our boundary conditions and thus not actually a continuum, may still be used to *interpolate* the actual, continuous energy distribution of an unbound particle. So, with the proper treatment, the coefficients a_n of Eq. (5.33) corresponding to positive eigenenergies of $\hat{H}_0$ *could* be used to estimate the actual energy distribution quite precisely. As this is slightly cumbersome, we will not do so here.

5.4 'Dynamics' with Two Particles

While most examples we have been playing with have remained within the safe confines of a single particle in one dimension, we have allowed ourselves a couple of excursions. We will allow ourselves one more and consider two particles – however, not really with explicit time dependence this time. Actually, it's not even time.

With two particles in one dimension represented by means of a discretized spatial grid divided into n pieces, it is convenient to represent the wave function as a square matrix:

$$\Psi(x_1, x_2) \to \begin{pmatrix} \Psi(x_0, x_0) & \Psi(x_0, x_1) & \cdots & \Psi(x_0, x_n) \\ \Psi(x_1, x_0) & \Psi(x_1, x_1) & \cdots & \Psi(x_1, x_n) \\ \vdots & \vdots & \ddots & \vdots \\ \Psi(x_n, x_0) & \Psi(x_n, x_1) & \cdots & \Psi(x_n, x_n) \end{pmatrix}. \tag{5.44}$$

Admittedly, the notation is somewhat confusing here; on the left hand side, the index on the position variable x refers to the particle, while on the right hand side it refers to the grid points. In this matrix representation, the rows refer to particle 1 and the columns to particle 2. Correspondingly, operators acting only on particle 1 may be implemented as left multiplication by an $(n + 1) \times (n + 1)$ matrix, while operators acting only on particle 2 correspond to right multiplication with the Ψ matrix.

5.4.1 Exercise: Symmetries of the Ψ-Matrix

Suppose the spatial two-particle wave function matrix of Eq. (5.44) corresponds to two identical fermions. Which possible symmetry properties is the matrix required to fulfil?

Perhaps, as you were studying Fig. 4.5, you asked yourself, 'Where does the "exact" ground state wave function come from?' Well, you are about to find out. In fact, you are about to *do it*.

5.4.2 Exercise: The Two-Particle Ground State

Our starting point is, again, the system of Exercise 4.4.3. However, this time we will not impose any specific shape of the wave function, such as Eqs. (4.24) or 4.26. We only require it to be symmetric.[10]

Now, with Ψ represented as a matrix, Eq. (5.44), why can the action of the full Hamiltonian, Eq. (4.28), be written as

$$\hat{H}\Psi \rightarrow h\,\Psi + \Psi\,h^{\mathrm{T}} + W \odot \Psi? \tag{5.45}$$

Here h is the matrix approximation of the single-particle part of the Hamiltonian, W is a matrix with the interaction for each pair of points and $\odot$ refers to element-wise matrix multiplication – the *Hadamard* product, which we first addressed in Exercise 3.4.5(b).

Now, Eq. (5.45) is an excellent starting point for doing actual dynamics with two particles. Here, however, we will settle for finding the ground state – by propagating, in imaginary time, the procedure we first implemented in Exercise 3.4.3. In that exercise we used a propagator of exponential form, see Eq. (3.30). However, as the imaginary time procedure tends to be a rather friendly and stable method, numerically speaking, we may attempt to simply implement it using Euler's forward method:

$$\Psi(t + \Delta t) \approx \Psi(t) - \hat{H}\Psi(t)\,\Delta t/\hbar. \tag{5.46}$$

As before, if we renormalize our wave function at each 'time' step, as in Eq. (3.30), the energy estimate may be obtained via the norm of the wave function at the next 'time' step. However, with a different 'propagator', the energy formula comes out somewhat differently from what we used in Exercise 3.4.3. How?

Now, implement this method and see if you can arrive at the same wave function and energy as the one shown in Fig. 4.5(d). In order to make it work with Euler's forward method, which your maths teacher may have warned you about, make sure not to be too ambitious when it comes to numerics. You may get a reasonable energy estimate using a box of size $L = 10$ with $n + 1 = 128$ grid points and a 'time' step of $\Delta t = 10^{-3}$ in our usual units.

Also, try to construct the result shown in Fig. 4.6(b) – the lowest energy state for a spatially exchanged anti-symmetric wave function. In order to ensure that you don't stray from the anti-symmetric path, enforce anti-symmetry from the outset and at each 'time' step: $\Psi \rightarrow 0.5(\Psi - \Psi^{\mathrm{T}})$; replace the wave function matrix with half the difference between the matrix and its own transpose.

Hopefully, you found both energy estimates and wave functions to be in reasonable agreement with Figs. 4.5 and 4.6.

[10] Symmetric, $\Psi = \Psi^{\mathrm{T}}$, not Hermitian, $\Psi = \Psi^{\dagger}$.

5.5 The Adiabatic Theorem

Dynamical processes are usually more interesting when something actually happens – brought about by more or less abrupt changes. However, in some situations it could be interesting to bring about changes as *slowly* as possible. We find one example of such in the realm of quantum computing – within what is called *adiabatic quantum computing*, which will be the topic of Section 6.6. Such approaches typically set out to solve complex optimization problems by initializing a quantum system in the ground state of a simple Hamiltonian and then slowly changing this Hamiltonian into something more complex.

5.5.1 Exercise: Adiabatic Evolution

(a) The generic Hamiltonian of a two-state system may be written as in Eq. (5.3); we repeat it here:

$$H = \frac{1}{2}\begin{pmatrix} -\epsilon & W \\ W^* & \epsilon \end{pmatrix}. \tag{5.47}$$

Contrary to the context of Eq. (5.3), we allow both the diagonal elements $\pm\epsilon/2$ and the couplings W to depend on time here. Show that, unless $W = 0$, the eigenenergies of this Hamiltonian will never coincide. This is illustrated in Fig. 5.3.

The same conclusion holds for any discrete set of eigenstates with more than two states as well.

(b) In Exercise 5.3.2 we used the time-independent part of the Hamiltonian in order to construct an orthonormal spectral basis. Suppose now that we instead use the eigenstates of the full, time-dependent Hamiltonian:

$$\hat{H}(t)\,\varphi_n(x;t) = \varepsilon_n(t)\,\varphi_n(x;t). \tag{5.48}$$

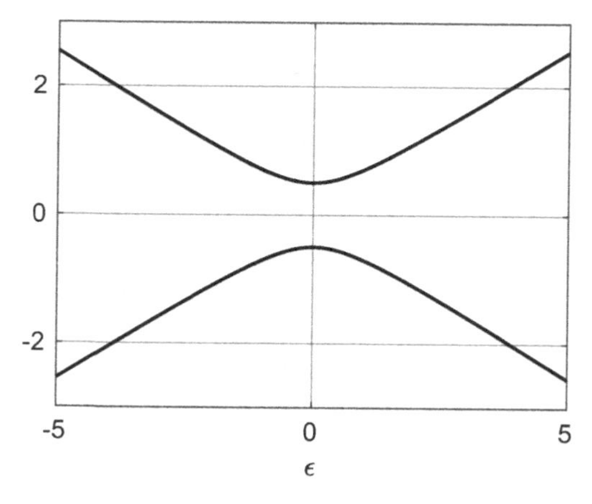

Figure 5.3 The eigenenergies of the two-state Hamiltonian in Eq. (5.47) as a function of the magnitude ϵ of the diagonal elements. The quantities are given in units of $|W|$. Do note the similarity with Fig. 3.5.

As before, we assume the spectrum of the Hamiltonian to be countable here. The orthonormal eigenfunctions $\varphi_n(x;t)$ can be set to vary continuously and smoothly in time – provided that the Hamiltonian does so.

Let our wave function be expanded in these time-dependent basis functions:

$$\Psi(x;t) = \sum_n a_n(t)\, \varphi_n(x;t). \tag{5.49}$$

Show that the state vector $\mathbf{a}(t) = (a_1, a_2, \ldots)^{\mathrm{T}}$, according to the Schrödinger equation, fulfils

$$i\hbar \frac{d}{dt}\mathbf{a}(t) = [D - iC]\,\mathbf{a}(t), \tag{5.50}$$

with the matrices

$$D = \begin{pmatrix} \varepsilon_0(t) & 0 & 0 & \\ 0 & \varepsilon_1(t) & 0 & \cdots \\ 0 & 0 & \varepsilon_2(t) & \\ & \vdots & & \ddots \end{pmatrix}, \tag{5.51}$$

$$C = \hbar \begin{pmatrix} 0 & \left\langle \varphi_0 \middle| \frac{\partial \varphi_1}{\partial t} \right\rangle & \left\langle \varphi_0 \middle| \frac{\partial \varphi_2}{\partial t} \right\rangle & \\ \left\langle \varphi_1 \middle| \frac{\partial \varphi_0}{\partial t} \right\rangle & 0 & \left\langle \varphi_1 \middle| \frac{\partial \varphi_2}{\partial t} \right\rangle & \cdots \\ \left\langle \varphi_2 \middle| \frac{\partial \varphi_0}{\partial t} \right\rangle & \left\langle \varphi_2 \middle| \frac{\partial \varphi_1}{\partial t} \right\rangle & 0 & \\ & \vdots & & \ddots \end{pmatrix}. \tag{5.52}$$

By the way, is the above effective Hamiltonian matrix, $D - iC$, Hermitian?

(c) Suppose now that our Hamiltonian $\hat{H}(t)$ depends very weakly on time; it changes very slowly. In that case, a system which starts out at time $t = 0$ in eigenstate number n tends to remain in eigenstate number n – although the Hamiltonian has changed.

How can we draw this conclusion from the above relations?

The result pertaining to part (a) of the exercise above, which is attributable to John von Neumann and Eugene Wigner, is called the *non-crossing rule*, and the result of (c) is coined the *adiabatic theorem*. Its name is motivated by thermodynamics; an adiabatic process is one that happens without heat exchange. In a quantum context, it simply refers to a process that proceeds slowly enough for no transition to other quantum states to occur; there is no exchange of population between the eigenstates. This applies to *discretized* time-dependent eigenstates; it does not hold for states belonging to any continuous part of the spectrum.

Simply put, the adiabatic theorem holds because when $\hat{H}(t)$ changes slowly, so do all the time-dependent eigenstates $\varphi_n(x;t)$ and $\langle \varphi_m(t) | \partial \varphi_n(t)/\partial t \rangle \approx 0$. With C approximately equal to the zero matrix, very little transition between states will occur. And, since none of the eigenenergies will coincide at any time, we will remain in the state with the nth lowest eigenenergy if that's where we started out. In particular, if we start out in the ground state of some initial Hamiltonian, we will end up in the ground state of the final one – if our time-dependent Hamiltonian has changed slowly and continuously between the two.

Although eigenenergies do not cross, they may come close; they may feature so-called *avoided crossings*, as illustrated in Fig. 5.3. Transitions between states are more likely to occur in these cases. This can be understood from the fact that the off-diagonal matrix elements of C in Eq. (5.52) may be recast as

$$\left\langle \varphi_m \left| \frac{\partial \varphi_n}{\partial t} \right. \right\rangle = \frac{\langle \varphi_m | \frac{\partial H(t)}{\partial t} | \varphi_n \rangle}{\varepsilon_n(t) - \varepsilon_m(t)}, \quad m \neq n. \tag{5.53}$$

From this we see that these coupling elements, which may shift population between eigenstates, are large not only when the time derivative of the Hamiltonian is large, as discussed above, but also when eigenenergies come close, $\varepsilon_m(t) \approx \varepsilon_n(t)$ – when they feature avoided crossings.

Enough talk, let's see if the adiabatic theorem actually holds for a specific example.

5.5.2 Exercise: A Slowly Varying Potential

Once again, we revisit the harmonic oscillator, which we first saw in Exercise 3.1.4. As before, we set $k = 1$. However, this time we shake it about a bit; we take our time-dependent Hamiltonian to be

$$\hat{H}(t) = \frac{\hat{p}^2}{2m} + V(x - f(t)) \tag{5.54}$$

with a time-dependent translation $f(t)$ in the form of Eq. (5.29):

$$f(t) = \frac{8}{3^{3/2}} \sin^2\left(\frac{1}{2}\omega t\right) \sin(\omega t), \quad t \in [0, 2\pi/\omega]. \tag{5.55}$$

The factor $8/3^{3/2}$ ensures that the maximal displacement is one length unit. The time-dependent potential is illustrated in Fig. 5.4.

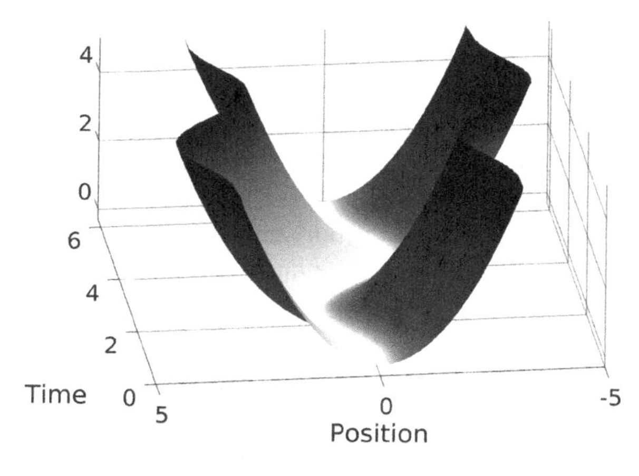

Figure 5.4 A harmonic potential is slowly shaken, first to the right and then to the left, before it is restored to its original position. The time is given in units of $1/\omega$.

According to the above discussion, if we start out in the ground state at $t = 0$, $\Psi(t = 0) = \varphi_0(t = 0)$, we should remain in the ground state at the final time $T = 2\pi/\omega$, provided that we shake slowly enough. In other words, ω must be low enough. Now, solve the Schrödinger equation for this system and try to determine just how low ω must be in order to ensure that the population in the initial state, the ground state population P_0, remains at least 99%:

$$P_0 = |\langle\Psi(t = 0)|\Psi(T)\rangle|^2 \geq 0.99. \tag{5.56}$$

Diagonalize your numerical approximation of the initial Hamiltonian in order to obtain your initial state, the normalized ground state.

Perhaps the most straightforward way to implement the time evolution would be the split-operator technique, Eq. (5.25), with $\hat{A} = \hat{p}^2/2m$ and $\hat{B} = V(x - \overline{f(t)})$, where

$$\overline{f(t)} \approx f(t + \Delta t/2) \tag{5.57}$$

is the time-averaged displacement. This is convenient since $\hat{A}$ is time independent and $\hat{B}$ is diagonal. Since we will run this calculation several times with various ω values, it would be more convenient to fix the number of time steps N_t, rather than the size of the numerical time step Δt. This ensures that each run takes the same amount of time and allows us to adapt the time step to the actual dynamics, $\Delta t = 2\pi/(\omega N_t)$.

You could also accept the challenge of implementing the evolution as formulated in Eq. (5.50) and solve it as an ODE, using Eq. (5.37) for instance. In doing so you can exploit the fact that, in this case, the eigenenergies are time independent and the elements of the C-matrix, Eq. (5.52), can be rewritten in terms of $\hat{p}$-couplings between the eigenstates at $t = 0$. As in Exercise 5.3.1, this matrix, in turn, may be written as a matrix product with the basis states and a matrix approximating the momentum operator – multiplied by the appropriate time-dependent factor.

While the first of the two alternative methods may be the most convenient one, it is always reassuring to check that different implementations yield the same result.

In order to determine how fast you can shake your potential and yet keep your ground state population in the end, you can start with a very low ω, rerun your calculation repeatedly while increasing ω in small steps, and see how far you get before you violate Ineq. (5.56). You could start at $\omega = 0.05$ and use the same or a smaller value for the increment. However, it may be more interesting to make a plot of P_0 as a function of ω. If you do this for ω up to, say, 2, you will find that this function is far from monotonic.

In addition to running your simulation starting out in the ground state, also try to start from some excited state and see how the probability of remaining in that state depends on ω.

In Exercise 6.6.1 we will try to illustrate how adiabatic evolution may be used as an optimization technique. When doing so, you will want to 'shake' your system slowly enough to keep your ground state population with high fidelity, while not wasting time progressing more slowly than necessary.

6 Quantum Technology and Applications

Quantum physics is the framework we use to understand the fundamental building blocks of matter; so anything that involves matter, also involves quantum physics in some sense. Thus, little is more *applied* than this. However, from a more practical point of view, quantum applications have long since become an integral part of our everyday lives. In addition to shifting our way of understanding the material world, our growing understanding of the micro world has been a game changer within several technological areas. As examples of such areas, we could mention medicine and diagnostics, computers and integrated circuits, solar cells, chemical and pharmaceutical industries, metrology – the art of making precise measurements, nuclear energy production and, sadly, the weapons industry.

Entering into how quantum physics is applied within all of these fields would be too immense an endeavour. We will, however, get a feel for a few applications which are particularly easily related to the quantum traits we have already seen – such as tunnelling, quantization and spin. Moreover, a large portion of this chapter is devoted to the emerging field of *quantum information technology*.

6.1 Scanning Tunnelling Microscopy

As mentioned, *tunnelling* may be used in resolving incredibly small structures since it depends quite sensitively on geometrical factors. This fact is exploited in a *scanning tunnelling microscope* – a microscope capable of producing images such as the one we saw in Fig. 2.5 and in Fig. 5.1(b). We will here try to outline how this comes about.

We start by assuming that we are dealing with a free particle with a definite energy. With this, we may assume our wave function to be of the form of Eq. (2.38):

$$\Psi(x;t) = e^{-i\omega t}\psi(x), \tag{6.1}$$

where ω is the energy ε divided by $\hbar$, $\omega = \varepsilon/\hbar$, and $\psi(x)$ is a solution of the time-independent Schrödinger equation, Eq. (3.1), which we will obtain. Although we didn't emphasize this when we first encountered this notion, in Exercise 2.5.1, the separation of Eq. (6.1) really only makes sense when dealing with bound states, which are subject to quantization. For a free particle, it is wrong in several ways. First of all, a particle with a definite energy exposed to no potential will also have a definite momentum. According to Werner Heisenberg, Ineq. (1.5), this can only be consistent with an entirely delocalized particle; if σ_p in Ineq. (1.5) approaches zero, σ_x will blow up. This, in turn, renders the wave function unnormalizable. Also, there is something fishy about the whole notion of using stationary solutions in order to understand something that really is dynamical. We are actually aiming at understanding time-dependent processes by solving a time-independent equation, Eq. (3.1) in this case.

Still, we will do so in the following. For three interrelated reasons: (1) while seemingly odd, this is indeed a very common approach to scattering problems in quantum physics, (2) it may actually lead to sensible results in the end, and (3) it *does* make sense when put in the right context.

6.1.1 Exercise: Tunnelling Revisited

Our starting point is the following: a one-dimensional particle with positive energy ε is exposed to a potential $V(x)$ which is only supported for x between 0 and D; for any x value outside the interval $[0, D]$, the potential energy is zero. The situation is illustrated in Fig. 6.1.

(a) Explain why the solution of the time-independent Schrödinger equation has the form

$$\psi(x) = Ae^{ikx} + Be^{-ikx} \quad \text{when} \quad x < 0, \tag{6.2a}$$

$$\psi(x) = Ce^{ik(x-D)} + Fe^{-ik(x-D)} \quad \text{when} \quad x > D \tag{6.2b}$$

in such a situation. What is k here?

As discussed above, this will lead to an unnormalizable wave function. So this is not really an admissible wave function as it does not produce any meaningful probability density. We couldn't really expect it to make sense either since we, after all, aim to study a scattering process – not a stationary solution or a *steady state*. The solution of Eqs. (6.2) is meaningful, however, if we we consider it a steady *current*.[1]

(b) If we combine each of the four terms in Eqs. (6.2) with the time-dependent factor of Eq. (6.1), $\exp(-\mathrm{i}\varepsilon t/\hbar) = \exp(-\mathrm{i}\omega t)$, we can identify each of them as a wave travelling either to the left or to the right. Now, if we insist that there is an incoming

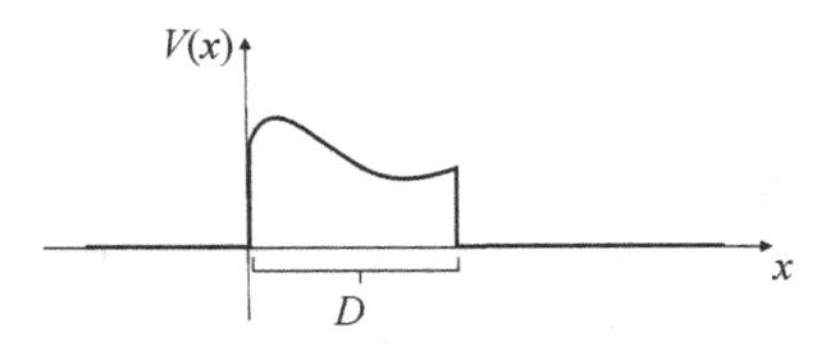

Figure 6.1 The potential under study in Exercise 6.1.1.

[1] The probability current corresponding to a wave function is determined as $j = \hbar/m \,\mathrm{Im}\, \Psi^* \mathrm{d}\Psi/\mathrm{d}x$. So for $\Psi(x;t) \sim \exp(\mathrm{i}(kx - \omega t))$, the current is proportional to k and independent of both position and time.

current from the left, but nothing from the right, one of the coefficients must be set to zero.

Which one is that? And which one of the three remaining terms corresponds to an incoming wave from the left?

In Exercise 2.4.1 we defined a reflection and a transmission probability, Eqs. (2.32). In our present situation, in which we are dealing with a steady current rather than a dynamic wave function, we define relative reflection and transmission *rates*. We define them as the squared modulus of the amplitude of the reflected and the transmitted current, respectively, divided by the amplitude of the incoming current:[2]

$$R = \left|\frac{B}{A}\right|^2, \tag{6.3a}$$

$$T = \left|\frac{C}{A}\right|^2. \tag{6.3b}$$

These rates will depend fully on the energy ε at which the potential $V(x)$ that the incoming wave will scatter. In the special case that our potential is constant, $V(x) = V_0$ when $x \in [0, D]$, we may determine the transmission/tunnelling rate exactly by analytical means. One can show that

$$T = \frac{1}{1 + \frac{V_0^2}{4\varepsilon(V_0-\varepsilon)} \sinh^2(\alpha D)}, \tag{6.4}$$

where $\alpha = \sqrt{2m(V_0 - \varepsilon)}/\hbar$ and $\sinh x = (e^x - e^{-x})/2$ is the co-called *hyperbolic sine function*.

We will not do so here. Instead we will calculate T numerically – in a manner that allows for a general shape of the potential.

(c) Suppose now that we do have the correct solution for $\psi(x)$ for $x \in [0, D]$. How does this allow us determine the reflection and transmission rates? For simplicity, set $C = 1$.

Hint: Although we have given up on the notion that ψ should be normalizable, we should still insist that $\psi(x)$ and $\psi'(x)$ are continuous everywhere.

(d) Explain how the problem can be formulated as this first-order, linear ordinary differential equation:

$$\frac{\mathrm{d}}{\mathrm{d}x}\begin{pmatrix}\psi(x)\\ \varphi(x)\end{pmatrix} = \begin{pmatrix}0 & 1\\ \frac{2m}{\hbar^2}V(x) - k^2 & 0\end{pmatrix}\begin{pmatrix}\psi(x)\\ \varphi(x)\end{pmatrix} \tag{6.5}$$

where $\varphi(x) = \psi'(x)$. Why must we impose the 'initial' value condition

$$\begin{pmatrix}\psi(x = D)\\ \varphi(x = D)\end{pmatrix} = \begin{pmatrix}1\\ ik\end{pmatrix}? \tag{6.6}$$

[2] We hope that you forgive the fact that we use the same names as in Eqs. (2.32) although we are actually calculating relative probability *rates* in this case.

(e) Implement the solution to this initial value problem. Use your ODE solver of preference, impose the condition of Eq. (6.6), fix some value for the energy ε, which, in turn, fixes k, and solve Eq. (6.5) backwards in x. Use your numerical solution to determine $\psi(x=0)$ and $\psi'(x=0) = \varphi(x=0)$ and, finally, use this to determine the transmission rate T, Eq. (6.3b).

As your first case, take your potential to be constant, $V(x) = V_0$, between $x = 0$ and $x = D$. Choose your own values for D and V_0. Rerun your implementation for several values of ε and confirm that it gives the same energy dependence as does Eq. (6.4).

(f) Now, take your potential to have this linear form:

$$V(x) = \begin{cases} V_0 - ax, & x \in [0, D], \\ 0, & \text{otherwise}. \end{cases} \tag{6.7}$$

Keep your values for V_0 and D and choose a modest, positive value for a – one that keeps $V(D)$ well above zero.

For this potential, determine the transmission rate as a function of the energy ε for energy values up to V_0.

(g) Often, the transition rate has an energy dependence that resembles an exponential. Use a logarithmic y-axis when plotting $T(\varepsilon)$ to gauge to what extent this is the case with your parameters.

A so-called *semi-classical* approach called the WKB approximation[3] allows us to estimate the transition rate or probability as

$$T \sim \exp\left[-\frac{2}{\hbar}\int_a^b \sqrt{2m(V(x)-\varepsilon)}\,\mathrm{d}x\right], \tag{6.8}$$

where the integration interval $[a, b]$ extends over the classically forbidden region, the region in which $\varepsilon - V(x) < 0$. The relation symbol '$\sim$' is here taken to mean 'approximately proportional to', which, undeniably, is a rather imprecise notion. In any case, sometimes this is a decent approximation, sometimes it's not. Here we will not discuss how and when. But we will make use of it in the next exercise – after having checked its validity against our implementation from Exercise 6.1.1.

The aim of the following exercise is to shed some light on how tunnelling may be exploited to produce pictures such as the one shown in Fig. 2.5.

Suppose we want to chart the surface of some metal. Within the metal, the least bound electrons, the *conduction electrons*, roam freely at a more or less definite energy ε. While they are not attached to any particular atom or ion, the collective attraction of the atomic nuclei does confine even these electrons to the metal.

This changes slightly when we introduce a sharp needle carrying a voltage relative to the surface of the metal. The voltage introduced by the needle distorts the potential experienced by the conductance electrons in the vicinity of the metal surface – thus

[3] It takes its name from Gregor Wentzel, Hendrik Anthony Kramer and Léon Brillouin. The last two may be found in Fig. 1.3.

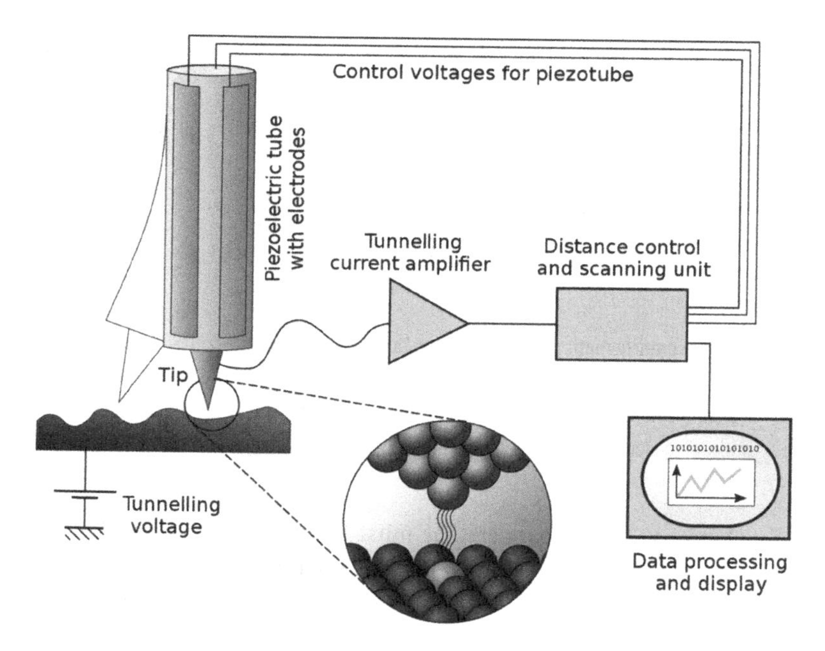

Figure 6.2 Illustration of the setup of a scanning tunnelling microscope. There is a voltage between the needle and the metal which permits free electrons in the metal to tunnel out, thus inducing a small current which can be measured. This provides information about the distance between the surface of the metal and the needle point.

allowing them to tunnel out. This is illustrated in Fig. 6.2. Correspondingly, a number of conductance electrons are able, through tunnelling, to escape the confines of the metal to a vacancy in the needle. Correspondingly, a small current is measured. The more likely tunnelling is to occur, the higher the current. As dictated, or rather *suggested*, by Eq. (6.8), the shorter the escape route, the more current is seen. In other words, this current is a measure of the distance between the surface and the needle point.

6.1.2 Exercise: The Shape of a Surface

Now, suppose that the surface of our metal plate is located in the yz-plane and our needle is pointing towards the negative x-direction. The distance between the needle and the surface is $d - f(y)$, where d is the average distance between the needle and the surface and $f(y)$ is the shape of the surface along the y-axis. Since the distance will deviate from d as you vary y, the current picked up by the needle will also depend on y. This scenario is illustrated in Fig. 6.3(a).

In a simplified picture we may model the potential energy felt by a conductance electron near the surface by the form of Eq. (6.7):

$$V(x;y) = \begin{cases} V_0 - \frac{eU}{d-f(y)}x, & x \in [0, d - f(y)], \\ 0, & \text{otherwise.} \end{cases} \tag{6.9}$$

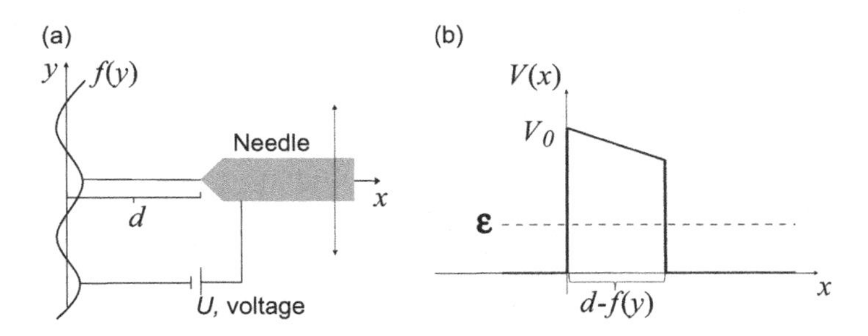

Figure 6.3 (a) A more schematic picture of the situation illustrated in Fig. 6.2. The shape of the surface along the y-axis follows the function $f(y)$, d is the average distance between the needle tip and the metal surface, and x is the distance from the metal surface for an escaping electron making its way towards the needle. (b) The potential that the escaping electron tunnels through, Eq. (6.9), in our simple model.

Here, V_0 is the barrier that must be climbed by a conductance electron in order to escape the metal[4] – if it were not for the tunneling possibility. The constant e is the elementary charge. The ratio between the *bias* voltage U and the distance between the needle point and the surface, $d - f(y)$, is the electric field experienced by the escaping electron. Thus, this term is identical to the one in Eq. (5.27). The potential is illustrated in Fig. 6.3(b).

Suppose now that we have continuously measured the tunnelling current while moving our needle along the y-axis. A set of fictitious results from such measurements can be found in a data file at www.cambridge.org/Selsto. Our aim is to use this measured current, $I(y)$, to determine the shape of the surface, $f(y)$. To this end, we will make use of Eq. (6.8).

In setting the parameters of our model, we will use units other than our usual, convenient ones – units that are more likely to be used in a laboratory. Set $d = 4$ Å, $V_0 = 4.5$ eV, $\varepsilon = 0.5$ eV and $U = 1$ V, where an *ångström*,[5] Å, equals 10^{-10} m and V is *volt*. Multiplying V by the elementary charge e renders the much applied energy unit, the *electronvolt*, eV.

(a) Convert all these parameters into atomic units. You may want to revisit Exercise 1.6.4 in case you need a reminder.
(b) For $f(y)$ values between -1 Å and 1 Å, calculate the actual tunnelling rates using your implementation from Exercise 6.1.1. To this end, set $D = d - f(y)$ and $a = eU/(d - f(y))$. Note that here a becomes y-dependent.
(c) Use Eq. (6.8) to predict the same transmission rate. Plot these predictions together with the ones you found in (b) using a logarithmic y-axis in order to gauge whether the predictions may reasonably be assumed to be proportional.

[4] This barrier height is referred to as the *work function*.

[5] The unit is assigned in tribute to the Swede Anders Jonas Ångström, a pioneer in the field of spectroscopy, to which the following section is dedicated.

(d) Now, assuming that you reached an affirmative conclusion to the question above, we may use Eq. (6.8) along with the assumption that the measured current $I(y)$ is proportional to the tunnelling rate $T(y)$ in order to determine the shape of the surface along the line in question. Under this assumption, it can be shown that

$$f(y) = \frac{3e\hbar U}{4\sqrt{2m}\left[(V_0 - \varepsilon)^{3/2} - (V_0 - \varepsilon - eU)^{3/2}\right]} \ln I(y) + C, \qquad (6.10)$$

where C is an unknown constant.

Now, do this analytical calculation for yourself.

(e) Finally, with the data for $I(y)$ from www.cambridge.org/Selsto, follow this procedure to estimate the shape $f(y)$. The constant C may be ignored, just subtract the mean f value from all the $f(y)$ values you arrive at.

So, by leading a needle very close to a surface, we can map out the shape of this surface by gauging the current that comes about via tunnelling. And we can cover the whole surface by moving our needle in both the y-and the z-directions. However, none of this is useful unless we are also able to control the position of the needle – with extreme precision. How does that come about? Here's a hint: *piezoelectricity*.

6.2 Spectroscopy

We have learned that the energy of the electrons confined to a nucleus in an atom is quantized; the energy can only assume one out of a set of discrete values. For molecules, the same also applies to vibrational and rotational energies. As a consequence, any substance has its own unique 'fingerprint' of allowed, discretized energies.

Although atoms and molecules tend to prefer the ground state, we may excite them. This can be done by heating, electromagnetic irradiation or by bombarding them with particles. This will populate several excited states. Afterwards, the excited system will relax towards the ground state – either directly or via intermediate states of lower energy. When this happens, the system disposes of the surplus energy by emitting light quanta – photons. The energy of these matches the energy difference between the two states in question. As a consequence, each photon comes out with a specific frequency and wavelength, see Eq. (1.4). When this wavelength falls within the interval between 385 nm and 765 nm,[6] more or less, it is visible to the human eye.

[6] nm – nanometre; 10^{-9} m.

6.2.1 Exercise: Emission Spectra of Hydrogen

In Exercise 3.1.3 we had a look at the eigenstates of the hydrogen atom. These energies turn out to have a very simple form – given by Eq. (3.4) in SI units. As we discussed in the context of Eq. (3.4), whenever an atom in an exited state, given by quantum number n_1, relaxes into one of lower energy with $n_2 < n_1$, a photon of energy

$$E_\gamma = \hbar\omega = hf = E_{n_1} - E_{n_2} = B\left(\frac{1}{n_2^2} - \frac{1}{n_1^2}\right) \tag{6.11}$$

is emitted. The corresponding wavelength is found from the relation $f\lambda = c$, where c is the speed of light.

(a) Now, only a few of these possible transitions correspond to wavelength in the visible spectrum. Which ones are they? How many are there? Which n_2 value(s) do they have?
(b) These particular wavelengths give rise to a *line spectrum* – a spectrum in which only a few colours are seen (see Fig. 6.4). Try to construct such a spectrum by plotting the wavelengths you found above. To make it more interesting, you may want to see if you can find an implementation that converts wavelength into colour – in the RGB[7] format, for instance. To this end, you may find Ref. [13] useful.

Although it is nowhere near as simple as the hydrogen spectrum, the helium atom, like any other atom or molecule, also has its own specific set of allowed electronic states. Thus, it also has its very specific spectral fingerprint.

6.2.2 Exercise: The Helium Spectrum

Among the resources at www.cambridge.org/Selsto you will find a file with data on the many bound states of the helium atom. It was retrieved from the NIST[8] Atomic Spectra Database [25]. The file consists of three columns. The third column lists all electronic energy levels in units of eV. This spectrum is shifted so that the ground state

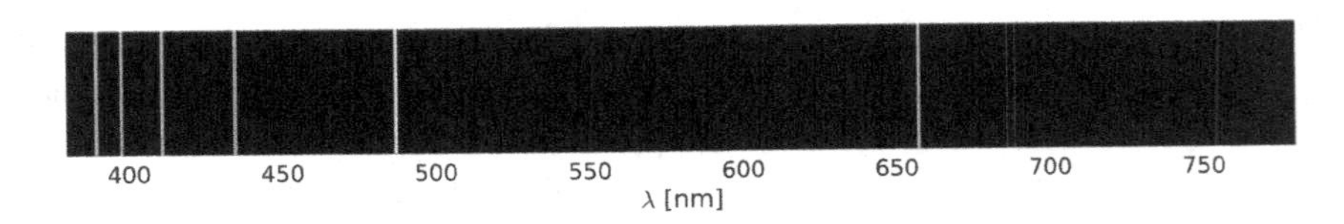

Figure 6.4 The lines indicate the wavelengths contained in light emerging from a gas of excited hydrogen atoms. They are called the Balmer series, named after the Swiss Johann Balmer, who arrived at Eq. (6.11), with $n_2 = 2$, on empirical grounds in 1885. Of course, such an illustration makes more sense in colour; a coloured version of this figure may be seen at www.cambridge.org/Selsto.

[7] RGB – red, green, blue.
[8] NIST – National Institute of Standards and Technology, USA.

energy is zero, which is not a problem for us since we will only be occupied with energy *differences*.

(a) Make an implementation that reads through all this data and compare pairs of energy levels to identify transitions corresponding to photon wavelengths that fall into the visible interval. Also, visualize this line spectrum as in Fig. 6.4.

How many lines does the spectrum contain? Some of them are quite close, right?

(b) Actually, the spectrum from (a) is far too inclusive, the spectrum that does in fact emerge from a gas of excited helium atoms has far fewer lines. There are several reasons for this.

First of all, helium is, just like the system we studied in Exercise 4.4.3, a two-particle system.[9] With a spin-independent Hamiltonian,[10] each eigenstate is the product of a spatial part and a spin part – either a triplet or the singlet. Correspondingly, the spatial wave function is either exchange anti-symmetric or symmetric, respectively. The emission of a photon will not change this symmetry. Consequently, the system will only undergo transitions within the same exchange symmetry.

Another restriction, or *selection rule*, which limits the number of possible transitions, is related to the fact that the photon has spin with $s = 1$. Since the total angular momentum, the sum of the 'ordinary' angular momentum and spin, is a conserved quantity in this process, the states involved in a one-photon transition must abide by this. To make a long story short, in this particular case the total angular momentum of the system has a well-defined quantum number L which *must* change by 1 in the case of a one-photon transition between one helium state and another.

In the data file from (a), the first column indicates if the state is a spin singlet, '1', or a spin triplet, '3'. The second column is the L quantum number.

Redo part (a) with these additional constraints implemented. What does your helium emission spectrum look like now?

(c) When white light, which contains components of all possible wavelengths and colours, passes through a gas, the opposite process of that which produces a line spectrum takes place. Where a line spectrum, or an *emission spectrum*, emerges when a gas of excited atoms and molecules emits light of specific wavelengths, an *absorption spectrum* comes about when a gas, with atoms and molecules predominantly in their ground states, absorbs light of specific wavelengths, thus leaving dark spots in the spectrum of the remaining light. Figure 6.5 displays one such absorption spectrum. It corresponds to sunlight that has passed through both the Sun's and the Earth's atmosphere.

Can you identify traces of hydrogen and helium in this absorption spectrum?

---

[9] If we include the nucleus, it actually has three particles. It is, however, admissible to consider the nucleus as a massive particle at rest – a particle that sets up the potential in which the two electrons are confined.

[10] Strictly speaking, it is more accurate to say that the Hamiltonian has a very weak spin dependence; spin is not entirely absent.

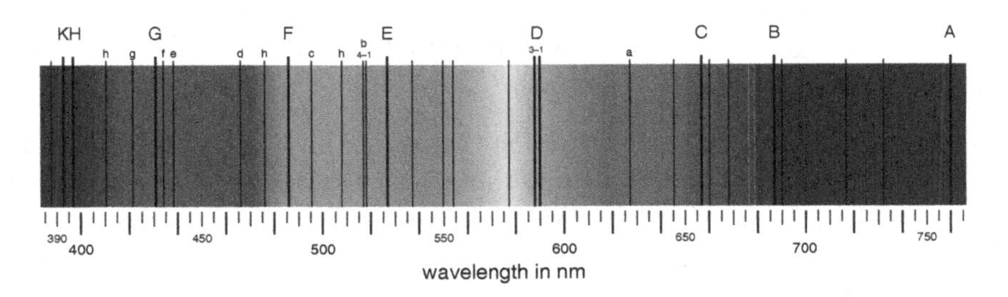

Figure 6.5 Since atoms and molecules in the atmospheres of the Sun and the Earth absorb light at certain frequencies, the sunlight that reaches us at the surface of the Earth has some 'holes' in its spectrum. These missing lines are called *Fraunhofer lines*, named after the German physicist Joseph von Fraunhofer. You may find the colour version of this figure at www.cambridge.org/Selsto more informative.

Hopefully, these examples serve to illustrate how unknown substances can be identified by investigating the light that they absorb or emit. Such techniques are referred to as *spectroscopy*. More sophisticated spectroscopic schemes than the ones outlined here do exist. And spectroscopic analysis is certainly not limited to the visual part of the electromagnetic spectrum.

By using spectroscopic techniques, even the composition of distant astronomical objects may be determined. And, albeit somewhat less exotic, spectroscopy has found several terrestrial applications, for instance within medical imaging.

The next topic has also found such imaging applications.

6.3 Nuclear Magnetic Resonance

We have learned that the particles that make up matter, the *fermions*, have half-integer spins. This also applies to the protons and neutrons – the building blocks of atomic nuclei. Thus, since nuclei may have a non-zero spin, they may also be manipulated by magnetic fields – in the same manner as we did in Section 5.1. By exposing matter to oscillating magnetic fields, atomic nuclei will respond in a manner that depends strongly on the frequency of the field. This may be exploited in order to image what we cannot see directly, such as tissue of a specific type inside a patient's body; *magnetic resonance imaging*, MRI, has long since become a standard tool in hospitals.

6.3.1 Exercise: Spin Flipping – On and Off Resonance

We now return to our system of a spin-1/2 particle with a static magnetic field in the z-direction and an oscillating one in the x-direction. However, this time our spin-1/2 particle will be a *proton*. Equation (4.15) still applies, but this time the charge is the positive elementary charge, the g-factor is 5.5857, and the proton mass is 1836 times greater than the electron mass.

We take the static magnetic field to point along the z-axis – so that the diagonal of the Hamiltonian matrix still coincides with Eq. (5.8) with positive ϵ.

(a) Show that, for the normalized spinor $\chi = (a(t), b(t))^T$, the time-dependent expectation value of the magnetic dipole, Eq. (5.2), is

$$\langle \boldsymbol{\mu} \rangle = \langle \chi | \hat{\boldsymbol{\mu}} | \chi \rangle = \frac{gq\hbar}{2m} \left[\mathrm{Re}\,(ab^*), -\mathrm{Im}\,(ab^*), |a^2| - 1/2 \right]. \tag{6.12}$$

Assume this time-dependent magnetic moment generates electromagnetic radiation in the same manner as a classical magnetic dipole. Moreover, for simplicity, limit your attention to the x-component of the magnetic moment expectation value. For a classical magnetic dipole oscillating in the x-direction, the intensity of the emitted radiation at distance r and angle θ relative to the x-axis is

$$I = \frac{m_0}{16\pi^2 c^3} \frac{\sin^2\theta}{r^2} \overline{\left[\frac{\mathrm{d}^2}{\mathrm{d}t^2} \langle \mu_x(t) \rangle \right]^2}, \tag{6.13}$$

where the angle brackets $\langle \cdot \rangle$ denote expectation value, as usual, and the bar is taken to indicate time averaging over the duration T of the interaction:

$$\overline{f(t)} = \frac{1}{T} \int_0^T f(t')\,\mathrm{d}t'. \tag{6.14}$$

The constant m_0 is the permeability of free space;[11] in SI units it takes the value $m_0 = 1.257 \cdot 10^{-6}$ N s^2/C^2.

(b) In order to simplify further, apply the rotating wave approximation, which you got to know in Exercise 5.1.3, in the following.

For a static magnetic field of magnitude B_z pointing along the z-axis, and an oscillating magnetic field in the x-direction oscillating at angular frequency ω, $B_x(t) = B_0 \sin(\omega t)$, identify the parameters δ and Ω that enter into the Hamiltonian of the rotating wave approximation, Eq. (5.15).

(c) Since the rotating wave Hamiltonian is time independent, the time evolution that it dictates may be determined by direct application of Eq. (2.19). Verify that with the adequate parameters Eq. (5.5) gives

$$\exp[-\mathrm{i}H_{\mathrm{RWA}}t/\hbar] = \cos(\Omega_{\mathrm{G}}t/2)\, I_2 - \mathrm{i}\frac{1}{\Omega_{\mathrm{G}}} \sin(\Omega_{\mathrm{G}}t/2) \left(\Omega\sigma_y - \delta\sigma_z \right). \tag{6.15}$$

Do note the distinction between the Rabi frequency Ω, which is real here, and the *generalized* Rabi frequency Ω_{G}, Eq. (5.18).

(d) With the spin-up state as our initial state, show that the time evolution of the spin-up and the spin-down amplitudes, a and b, respectively, reads

$$a(t) = e^{\mathrm{i}\omega t/2} \left(\cos(\Omega_{\mathrm{G}}t/2) + \mathrm{i}\frac{\delta}{\Omega_{\mathrm{G}}} \sin(\Omega_{\mathrm{G}}t/2) \right), \tag{6.16a}$$

$$b(t) = e^{-\mathrm{i}\omega t/2} \frac{\Omega}{\Omega_{\mathrm{G}}} \sin(\Omega_{\mathrm{G}}t/2). \tag{6.16b}$$

[11] Usually, this fundamental constant is denoted by μ_0. Here, however, we call it m_0 in order to avoid confusion with the magnetic moment.

If the factors $\exp(\pm \mathrm{i}\omega t/2)$ appear somewhat mysterious, do remember that you need to transform back to the original frame according to Eq. (5.16).

(e) Make an implementation that estimates, numerically, and plots the time average of $[\mathrm{d}^2/\mathrm{d}t^2 \langle \mu_x(t) \rangle]^2$ as a function of the driving frequency $f = \omega/2\pi$. According to Eq. (6.13), the intensity of the electromagnetic signal emitted from the spin-flipping proton is proportional to this quantity. You may find Eqs. (2.11a) and (1.16a) useful. Let the frequency interval span from 100 MHz to 300 MHz and fix the duration of the interaction at 100 optical cycles of the driving field, $T = 100/f$. Fix the static magnetic field at $B_z = 5$ T and let the amplitude of the oscillating field be considerably lower: $B_0 = 0.01$ T. One tesla, the SI unit of magnetic field strength, is the same as kg/C s.

(f) Does any input frequency f generate a stronger output signal than any other frequency? Which frequency is that?

(g) Try increasing T and decreasing B_0 to gauge how the strength of the emitted radiation from the spin-flipping proton is affected.

(h) Is the rotating wave approximation really justified here?

Hopefully you found that the spin system responded a lot more strongly to oscillating magnetic fields at a specific angular frequency ω – namely the one for which the detuning δ is zero. Perhaps the output signal looked something like the sketch in Fig. 6.6? This phenomenon, which is seen and used in virtually all systems that involve oscillations and waves – such as a swing on a playground or a radio signal – is called *resonance*.

The answer to the last question in Exercise 6.3.1, the one about the applicability of the rotating wave approximation, is *no*; while the Rabi coupling Ω is, indeed, small, we are guilty of applying the rotating wave approximation at frequencies rather far away from resonance. However, a more proper description of the dynamics would still produce the same feature. The signal is many orders of magnitude stronger when the frequency of the input signal matches the resonant transitions between the two spin

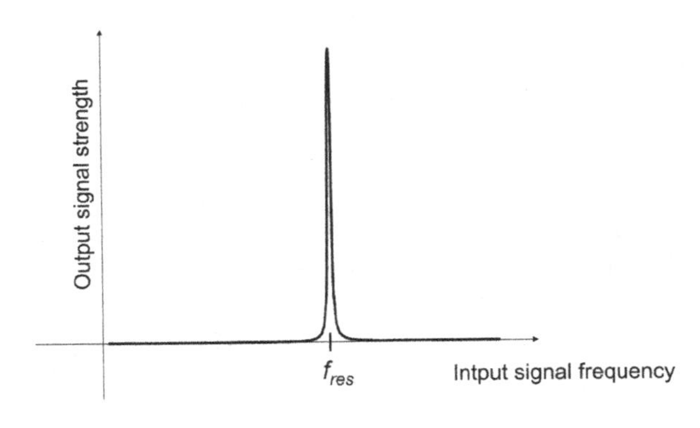

Figure 6.6 The phenomenon of resonance – how some physical system may respond very strongly to an oscillating input signal at a specific frequency.

states – when δ is very close to zero. It is this region we are interested in, and it is in this very frequency regime that the rotating wave approximation applies.

The fact that a nucleus, be it the proton in a hydrogen atom or the nucleus of a larger atom, responds so strongly to a very particular frequency is called *nuclear magnetic resonance*. In hospitals, this phenomenon is exploited in a technique called *magnetic resonance imaging* (see Fig. 6.7). Since the origin of the emitted signal can be determined rather accurately, see Eq. (6.13), the location of specific tissue inside a living body may be imaged. This can be done with much more sophistication and in a far less invasive manner than X-ray imaging, for instance.

With all constants in Eqs. (6.12) and (6.13) in place, you may find that the intensity of the field induced by the spin-flipping proton isn't very high. However, since there are many, many protons in the molecules that make up biological tissue, the signal is certainly one we are able to measure. But isn't that also a problem? Since all biological tissue is full of hydrogen atoms, how can you tell one type of tissue from any other? Wouldn't all the protons respond resonantly to exactly the same frequency?

Fortunately, the answer to this question is *no* – due to so-called *chemical shielding*. The electrons of the molecule, which carry both charge and spin, will shield the nucleus a bit – effectively reducing the magnetic field experienced by the nucleus. This, in turn, shifts the resonance frequency a little bit downwards. Since different molecules have different electron distributions, this shift differs from molecule to molecule. While not very large, the shift *is* large enough for us to distinguish between different kinds of tissue.

Gauging the emitted field during exposure to an oscillating magnetic field with a certain frequency is not the only way to gather information about what's inside a body. More information may be extracted by using a pulsed oscillating field

Figure 6.7 Magnetic resonance imaging, MRI – a direct application of nuclear magnetic resonance – has long since become a standard imaging technique in hospitals. Here a patient is being positioned for MR study of the head and abdomen.

instead – combined with Fourier analysis. Also, the relaxation of aligned nuclear spins to so-called *thermal equilibrium* after exposure to the magnetic fields carries additional useful information.

6.4 The Building Blocks of Quantum Computing

From Section 1.3 you may remember the infamous *curse of dimensionality*. Actually, this curse was the motivation for introducing the concept of a *quantum computer*. One of its advocates was the 1965 Nobel laureate Richard Feynman. He argued that, since the complexity of quantum physical systems grows exponentially with the number of particles involved, a numerical study of such systems is best done on a computer with the same capacity [15].

In the following we will illustrate how this works with spin-1/2 particles. We will also use spin-1/2 particles to introduce the concepts of quantum bits, *qubits*, and *quantum gates*.[12]

Let's re-examine Eqs. (4.13) and (4.10) – and also include a general expression for a three-particle spin-1/2 state:

$$\chi_1 = a\,\chi_\uparrow + b\,\chi_\downarrow \tag{6.17a}$$

$$\chi_2 = a\,\chi_\uparrow^{(1)}\chi_\uparrow^{(2)} + b\,\chi_\uparrow^{(1)}\chi_\downarrow^{(2)} + c\,\chi_\downarrow^{(1)}\chi_\uparrow^{(2)} + d\,\chi_\downarrow^{(1)}\chi_\downarrow^{(2)} \tag{6.17b}$$

$$\begin{aligned}\chi_3 = {} & a\,\chi_\uparrow^{(1)}\chi_\uparrow^{(2)}\chi_\uparrow^{(3)} + b\,\chi_\uparrow^{(1)}\chi_\uparrow^{(2)}\chi_\downarrow^{(3)} + c\,\chi_\uparrow^{(1)}\chi_\downarrow^{(2)}\chi_\uparrow^{(3)} + d\,\chi_\uparrow^{(1)}\chi_\downarrow^{(2)}\chi_\downarrow^{(3)} \\ & + e\,\chi_\downarrow^{(1)}\chi_\uparrow^{(2)}\chi_\uparrow^{(3)} + f\,\chi_\downarrow^{(1)}\chi_\uparrow^{(2)}\chi_\downarrow^{(3)} + g\,\chi_\downarrow^{(1)}\chi_\downarrow^{(2)}\chi_\uparrow^{(3)} + h\,\chi_\downarrow^{(1)}\chi_\downarrow^{(2)}\chi_\downarrow^{(3)}\end{aligned} \tag{6.17c}$$

If we continue along this path, we will quickly run out of letters in the alphabet. Perhaps you see a tendency in the number of coefficients here – in the dimensionality of the spin wave functions?

6.4.1 Exercise: A Blessing of Dimensionality

(a) How many coefficients would you need to express the spin wave function for a system consisting of n spin-1/2 particles?

(b) Suppose you need 64 (usual) bits to represent a real number with reasonable accuracy on your computer. Suppose also that your computer memory is 16 GB.[13] Now, if you want to simulate a quantum system consisting of several non-identical spin-1/2 particles, how many such particles would you be able to encompass in your computer's memory without hitting the roof?

(c) Suppose now that we turn the tables. We use our set of spin-1/2 particles to encode and process information – rather than trying to simulate these spins. How, then, is the above good news?

[12] Spoiler: this is just another word for *propagator*.

[13] GB – gigabyte, where one byte is eight bits.

The answer to the last question above may not be obvious. Perhaps it is more obvious that the answer to the first question is 2^n. This is the curse of dimensionality revisited; the complexity grows exponentially with the number of constituent parts of our quantum system. We could turn this into an advantage. Suppose we have access to n spin-1/2 particles, and that we can manipulate them as we see fit. This would mean that we are able to handle an enormous content of information with a comparatively low number of two-state quantum systems. It shouldn't be hard to imagine that the potential is huge – if we are able to take advantage of this somehow. It should also be said that this is not trivial, finding efficient ways to encode classical information into an entangled quantum system is an active area of research.

The state of Eq. (6.17a) may be used as a quantum bit – a *qubit*. A qubit is a linear combination of two orthogonal states of a two-state quantum system – *any* two-state quantum system. As mentioned towards the end of Section 5.1, the spin projection of a spin-1/2 particle is an example of such a system. Another example of such a two-dimensional system is the polarization of a photon – left/right vs. up/down, for instance.

Yet another possible realization could be the two lowest bound states of a confined, quantized quantum system, such as an atom, an ion or a *quantum dot*, see Fig. 5.1. All of these systems could act as qubits – under the condition that the dynamics could be rigged, somehow, so that the other bound states and the continuum could be ignored. See Fig. 4.4 for a schematic illustration of how the core of a *trapped ion quantum computer* could look. Specific ions are manipulated by hitting them with tailored laser beams – beams that may induce jumps between the two bound states in question.

In general, it is customary to label the two states analogously to classical bits – by '0' and '1'. Moreover, they are typically written using Dirac notation – with brackets: $|0\rangle$ and $|1\rangle$. Correspondingly, a general qubit reads

$$|\Psi\rangle = a|0\rangle + b|1\rangle, \tag{6.18}$$

where, as for any wave function/quantum state, we insist that it is normalized:

$$\langle\Psi|\Psi\rangle = |a|^2 + |b|^2 = 1. \tag{6.19}$$

At the risk of being excessively explicit, the only difference between Eq. (6.18) and Eq. (6.17a) is the labelling, we have just substituted '$\chi_\uparrow$' by '$|0\rangle$' and '$\chi_\downarrow$' by '$|1\rangle$'. Thus, a system consisting of n non-identical spin-1/2 particles could always serve as a set of qubits, and any other set of qubits could be perceived as such. And, as before, the two states may be represented as simple vectors in $\mathbb{C}^2$:

$$|0\rangle = \begin{pmatrix} 1 \\ 0 \end{pmatrix} \quad \text{and} \quad |1\rangle = \begin{pmatrix} 0 \\ 1 \end{pmatrix}, \tag{6.20}$$

so that a general qubit may also be written[14]

$$|\Psi\rangle = \begin{pmatrix} a \\ b \end{pmatrix} \tag{6.21}$$

as in Eq. (4.13).

[14] We apologize for the abundance of different names for what is basically the same thing.

6.4.2 Exercise: The Qubit

From Eq. (6.18) it would seem that it would take four real numbers to specify a qubit:

$$|\Psi\rangle = (\mathrm{Re}\, a + \mathrm{i}\,\mathrm{Im}\, a)\,|0\rangle + (\mathrm{Re}\, b + \mathrm{i}\,\mathrm{Im}\, b)\,|1\rangle = \begin{pmatrix} \mathrm{Re}\, a + \mathrm{i}\,\mathrm{Im}\, a \\ \mathrm{Re}\, b + \mathrm{i}\,\mathrm{Im}\, b \end{pmatrix}. \tag{6.22}$$

However, it is actually just two. Why?

Maybe it becomes clearer if we rewrite Eq. (6.22) as

$$\begin{pmatrix} a \\ b \end{pmatrix} = e^{\mathrm{i}\varphi_a} \begin{pmatrix} |a| \\ |b| e^{\mathrm{i}(\varphi_b - \varphi_a)} \end{pmatrix}, \tag{6.23}$$

where $|a|$ is the modulus of a and φ_a is its argument.

Does the fact that a single qubit is fixed by two *real* numbers rather than by two complex ones have severe consequences for the conclusion drawn at the end of Exercise 6.4.1?[15]

The point here is that, as we learned in Exercise 1.6.3(c), any global phase, such as $\exp(\mathrm{i}\varphi_a)$, is immaterial; it does not change any physics or information content. Whenever the wave function, or the *state*, is used to calculate anything physical, such as an expectation value, $\langle\Psi|\hat{A}|\Psi\rangle$, it will cancel out. As for the magnitude of the coefficients, $|a|$ and $|b|$, these quantities are restricted by normalization, Eq. (6.19).

In the literature you will also see Eq. (6.23) expressed as

$$|\Psi\rangle = \cos\frac{\theta}{2}|0\rangle + e^{\mathrm{i}\phi}\sin\frac{\theta}{2}|1\rangle, \quad \text{where} \tag{6.24}$$
$$\tan\frac{\theta}{2} = \frac{|b|}{|a|} \quad \text{and} \quad \phi = \varphi_b - \varphi_a,$$

which motivates visualizing a single qubit as a point on a sphere of unit radius (see Fig. 6.8).

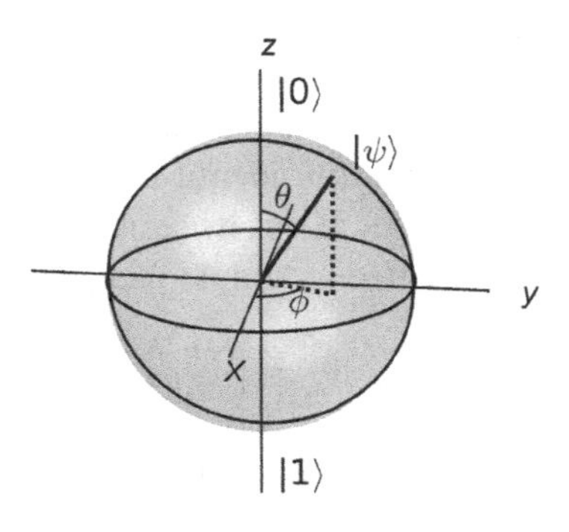

Figure 6.8 The Bloch sphere. This is a graphical representation of a general single qubit. The geometrical correspondence between these angles and the parameters in Eq. (6.23) is provided in Eq. (6.24).

[15] You may find this to be a bit of a leading question.

As in the case of 'traditional', classical information processing, we want to be able to manipulate qubits. We do so by imposing a sequence of operations on the various qubits that our quantum computer consists of. These gates may address individual qubits or several simultaneously. Clearly, the former is more easily dealt with than the latter. In any case, the manipulation comprises a sequence of operations of the form

$$|\Psi\rangle \to \hat{U}|\Psi\rangle, \tag{6.25}$$

where $\hat{U}$ is a linear operator.

6.4.3 Exercise: Quantum Gates and Propagators

In Section 2.3 we introduced the operator $\hat{U}$ that brings our initial quantum state $\Psi(t=0)$ into the final state at time t, $\Psi(t)$.

(a) Suppose the system at hand has a Hamiltonian with no explicit time dependence. In this case, $\hat{U}$ may formally be expressed in a particularly simple form. How?
(b) Suppose now that we are dealing with a single qubit only. Why is it then sufficient to learn how $|0\rangle$ and $|1\rangle$ are affected by the gate $\hat{U}$ in order to determine the effect on any qubit state?
(c) The NOT gate is a cornerstone in both 'ordinary', classical computing and quantum computing. It changes 0 to 1 and 1 to 0 – with or without brackets. With our single qubit expressed as in Eq. (6.21), what does $\hat{U}$ look like for this gate? Have you seen this matrix before?

Of course you have; the NOT gate is nothing but the Pauli matrix σ_x, which we encountered in Exercise 4.3.1. The other Pauli matrices are also frequently encountered in quantum computing – along with the *Hadamard gate*:

$$U_{\mathrm{H}} = \frac{1}{\sqrt{2}}\begin{pmatrix} 1 & 1 \\ 1 & -1 \end{pmatrix}. \tag{6.26}$$

6.4.4 Exercise: Quantum Gates are Unitary

A propagator or a quantum gate needs to preserve the norm of the quantum state it is acting on. This leads to the requirement that it must be unitary:

$$\hat{U}^\dagger = \hat{U}^{-1} \quad \Leftrightarrow \quad \hat{U}^\dagger \hat{U} = \hat{I}. \tag{6.27}$$

(a) Why is conservation of the norm crucial in quantum physics?
(b) Prove that Eq. (6.27) actually ensures that the norm of the state is conserved.
(c) As we have seen, in Eq. (2.19), when the Hamiltonian H is time independent the propagator may be written as $\hat{U} = \exp(-\mathrm{i}\hat{H}t/\hbar)$. How does this form ensure unitarity? What's the requirement?
(d) Check, by explicit calculations, that the Pauli matrices and the Hadamard gate are, in fact, all unitary.

We have seen that the Pauli matrices may be used to express the interaction between a spin-1/2 particle and a magnetic field. This, in turn, allows us to construct quantum gates on spin qubits by tailoring these magnetic fields. Specifically, in Exercise 4.3.1 we learned that the Pauli matrices are both Hermitian and their own inverses. Thus, they are, as we just saw, also unitary which, in turn, also means that they are admissible single-qubit quantum gates. Here we will see how they may be implemented on spin qubits.

6.4.5 Exercise: Pauli Rotations

This exercise is, to a large extent, a revisit of Exercise 5.1.1.

In quantum computers, gates are often implemented by subjecting qubits to piece-wise constant influences of some sort. Suppose a single qubit in the form of a spin-1/2 particle is exposed to a magnetic field pointing in the x-direction with a duration of T time units:

$$B_x(t) = \begin{cases} B_0, & t \in [0, T], \\ 0, & \text{otherwise}. \end{cases} \tag{6.28}$$

Then Eq. (2.19) applies during the time interval in question, and the Hamiltonian may be written

$$H = \frac{1}{2} W \sigma_x. \tag{6.29}$$

(a) By comparing with Eqs. (4.15) and (4.16), how does the coupling strength W relate to the strength B_0 of the magnetic field?

(b) Suppose we fix the strength of the magnetic field B_0 and, thus, also W, while the length of the pulse T may be tuned. Use Eq. (5.5) to tune T so that the propagator $\hat{U}$ coincides with the NOT gate.

In case you find the factor $-\mathrm{i}$ in the second term in Eq. (5.5) rather pesky, do remember the lesson learned in Exercises 1.6.3(c) and 6.4.2: a global phase factor does not matter.

(c) Verify that, by directing the magnetic field in the y- or the z-direction, the other Pauli matrices may also be constructed as quantum gates.

(d) Show that the Hadamard gate, Eq. (6.26), can be implemented by a Pauli rotation followed by the NOT gate.

Actually, it can also be implemented by first imposing a σ_z gate followed by the same Pauli rotation. Show this too.

It is useful to have an idea of how to manipulate single qubits. However, in order to implement meaningful quantum algorithms, we must also be able to mix them up. Useful quantum algorithms require that we can address multiple qubits simultaneously – that we can construct gates which involve two or more qubits.

This means that the particles must be brought to interact somehow. We introduced an example of such an interaction in Exercise 5.1.4.

Analogously to Eq. (5.20), any two-qubit state may be written as a linear combination of product states. We write $\chi_\uparrow^{(1)}\chi_\uparrow^{(2)}$ as $|00\rangle$ and so on, and, in this context, rewrite Eq. (5.20) as

$$|\Psi\rangle = a\,|00\rangle + b\,|01\rangle + c\,|10\rangle + d\,|11\rangle \to \begin{pmatrix} a \\ b \\ c \\ d \end{pmatrix}. \tag{6.30}$$

Two important two-qubit gates are the controlled-NOT (CNOT) gate and the SWAP gate. The latter simply turns the first qubit into the second and vice versa, $|10\rangle \to |01\rangle$ and so on. The CNOT gate flips the second qubit – but only if the first one is 1; the first bit, the *control bit*, remains unchanged.

6.4.6 Exercise: CNOT and SWAP

(a) With the vector representation of Eq. (6.30), what do the CNOT and SWAP matrices look like? Remember that the matrices corresponding to these linear transformations may be determined by working out how each of the four basis vectors is transformed.

(b) In the setup of Exercise 5.1.4, CNOT simply cannot be implemented. Why is that?
Hint: Remember the lesson learned in Exercises 2.6.2 and 5.1.4; it has to do with exchange symmetry.

(c) We could envisage a controlled switch off-gate – one that turns the second qubit into $|0\rangle$ if the first qubit, the control bit, is $|1\rangle$ and does nothing if the control bit is $|0\rangle$. You will never find such a gate in a quantum computer. Why is that?

(d) Now we will turn off the external magnetic fields in Eq. (5.21) completely, so that both ϵ and A are zero. Also, for simplicity, we set the spin–spin interaction strength $u = 1$, and $\hbar = 1$, as usual:

$$H = \begin{pmatrix} 1 & 0 & 0 & 0 \\ 0 & -1 & 2 & 0 \\ 0 & 2 & -1 & 0 \\ 0 & 0 & 0 & 1 \end{pmatrix}. \tag{6.31}$$

Suppose that we can manipulate our quantum system so that the spin–spin interaction is switched on only for a finite time T – for instance by bringing the particles close and then apart again. Then this Hamiltonian is able to implement the SWAP gate if we tune the duration T adequately – analogously to what we did in Exercise 6.4.5(c).

What is the minimal value for T that achieves this?

This could be done by analytical means. It may also be done numerically, however. One way of doing so is to vary the duration T in order to find the minimum of the function

$$C(T) = 1 - \left|\frac{1}{4}\mathrm{Tr}\left(U_{\mathrm{SWAP}}^\dagger U(T)\right)\right|^2 \tag{6.32}$$

where $U(T)$ is the actual gate for duration T, given by Eq. (2.19) with Eq. (6.31), and U_{SWAP} is the SWAP gate. The function $C(T)$ is zero if, and only if, our gate U is identical to the SWAP gate – up to a trivial phase factor. 'Tr $(\cdot)$' means that we should calculate the *trace* of the matrix in question, which simply is the sum of all the diagonal elements.

We introduced the concept of *entanglement* and Bell states in Section 4.2. And we played around with the entangled spin states of Eqs. (4.11). In Dirac notation these two states may be written:

$$|\Psi^+\rangle = \frac{1}{\sqrt{2}}\left(|01\rangle + |10\rangle\right), \tag{6.33a}$$

$$|\Psi^-\rangle = \frac{1}{\sqrt{2}}\left(|01\rangle - |10\rangle\right). \tag{6.33b}$$

These are maximally entangled states for which a readout/measurement necessarily would yield opposite results for the two qubits; if a measurement on the first qubit yields the result 0, the other one must be 1, unless you have tampered with it after measuring the first. The following two two-qubit states are also maximally entangled, but in this case a readout is destined to yield the same result for both qubits:

$$|\Phi^+\rangle = \frac{1}{\sqrt{2}}\left(|00\rangle + |11\rangle\right), \tag{6.34a}$$

$$|\Phi^-\rangle = \frac{1}{\sqrt{2}}\left(|00\rangle - |11\rangle\right). \tag{6.34b}$$

The four states of Eqs. (6.33) and (6.34) are known as the *Bell states*. As mentioned, they are named after John Stewart Bell, whose picture you can see in Fig. 6.9.

As we alluded to in Exercise 4.2.1(d) and saw in Exercise 6.4.1, entanglement of quantum states opens up a vast space for doing computations. The ability to construct quantum algorithms with an advantage over traditional, classical ones relies on this capacity. Certain schemes from quantum information theory, such as *superdense coding* and *quantum teleportation* for instance, make explicit use of the Bell states.

6.4.7 Exercise: Prepare Bell

A quantum computer typically starts out with just zeros as the initial state. In the two-qubit case, that would be the state $|00\rangle$.

Suppose that, starting out with $|00\rangle$ we apply first, the Hadamard gate to the first qubit, and second, the CNOT gate with the first bit as control bit. This will produce one of the Bell states. Which one is that?

Using CNOT, Hadamard and NOT, all four Bell states can be constructed from $|00\rangle$. Try to work out how this is achieved for the other three Bell states.

While we consistently have been alluding to spin-1/2 particles when addressing quantum bits and quantum gates, this is, as mentioned, just one out of several possible

Figure 6.9 John Stewart Bell (1928–1990) at CERN (*Conseil Européen pour la Recherche Nucléaire*), Geneva, Switzerland, in 1984. The 2022 Nobel Prize in Physics was awarded to Alain Aspect, John F. Clauser and Anton Zeilinger for conducting experiments based on his ideas. This, of course, is a clear recognition of Bell's work also.

realizations. It should also be mentioned that quantum information theory may very well be studied without regard to practicalities such as the physics behind the qubits or how the gates are actually implemented.

6.5 Quantum Protocols and Quantum Advantage

In the preceding section we introduced a few of the basic building blocks of quantum computing and quantum information processing. And in Exercise 6.4.1 we learned that the potential information content could be huge – even with just a few working qubits. In fact, with about 55 qubits[16] we would be able to process an information content beyond any present-day supercomputer.

And there is more: a quantum computer is parallel by nature. You may be familiar with parallel processing – the ability to chop a large computational job into chunks and distribute them to different processing units (CPUs). On a quantum computer, however, processing is parallel by nature. It is given directly by linearity; since the gate/propagator in Eq. (6.25) is linear, linear combination of initial states is preserved,

$$a|\psi_1\rangle + b|\psi_2\rangle \to \hat{U}\,(a|\psi_1\rangle + b|\psi_2\rangle) = a\hat{U}|\psi_1\rangle + b\hat{U}|\psi_2\rangle. \tag{6.35}$$

[16] That would be 55 noiseless, *logical*, qubits, mind you.

Correspondingly, we could prepare our initial state in a linear combination of *all* possible inputs and have the output for each and every one of them in a single calculation.

So the advent of working quantum computers is much anticipated – with a great deal of expectation and optimism. However, actually harnessing the huge advantage quantum computers seem to have over ordinary, classical ones, is not that straightforward. First of all, the collapse of the wave function prevents us from directly exploiting the quantum parallelism just mentioned; while the final state of an initial state with several inputs does contain all corresponding outputs, we can only read off one each time. Upon readout/measurement, the state will collapse – in a non-deterministic manner. To access different outputs, you would have to rerun your quantum program several times – leaving you, as Max Born taught us, with a distribution of answers, each one with its own frequency/probability. In order to acquire the output, or outputs, of interest, you may have to run your quantum calculation so many times that the quantum parallelism does not really offer any advantage.

The true potential in harnessing quantum parallelism and what we coined a *blessing of dimensionality* in Exercise 6.4.1 is more subtle. Any idea along the lines of 'let's run this computation on a quantum computer instead of a classical one in order to speed it up' is too naive. However, if we can orchestrate a quantum calculation so that only outputs of interest are read off in the end and all others come out with zero amplitude, there may be a speedup over classical computing – a *quantum advantage*. This, in turn, requires a different approach to programming, a quantum way of thinking about it.

There is a rich zoo of specific algorithms that succeed in exploiting the quantum advantage [23]. Some of these schemes serve mostly to illustrate the advantage of quantum computing from a rather academic point of view. Others, such as the *quantum Fourier transform* and *Shor's algorithm*, are likely to be very important when large and robust enough quantum computers are accessible. The problem with such schemes, however, is that they require a large number of noiseless qubits. It is hard to keep a quantum computer entirely isolated from interference from its surroundings. The quantum system and its environment tend to get entangled – with the consequence that the quantum system gradually loses its coherence property, and interference is ruled out. We will touch upon this issue in the next chapter.

There are quantum schemes, however, that are less sensitive to noise and *decoherence*.

Enough of these general considerations. The notion of a possible quantum advantage to computation and information processing makes a lot more sense if we know some specific examples. We will not indulge ourselves with any study of specific algorithms for quantum computing *per se*. We will, however, look at two illustrative quantum communication protocols. It is fair to say that the first one, although it has been realized experimentally, falls into the category of being mostly of academic interest. The second one, on the other hand, addresses *quantum key distribution*, for which commercial implementations are already offered.

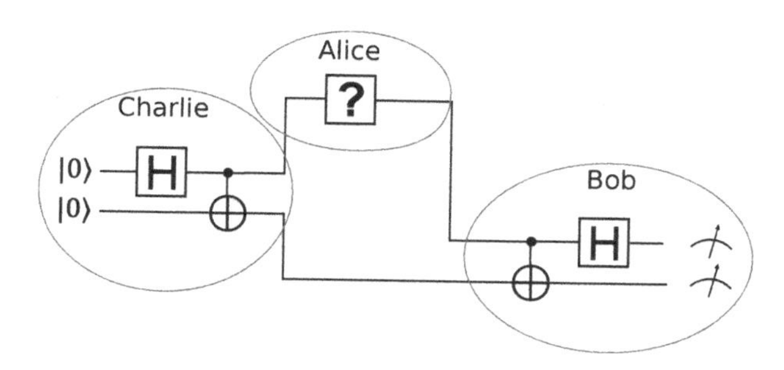

Figure 6.10 The superdense coding scheme. The two piecewise horizontal lines depict the two qubits. The H operator, acting on the first qubit, is the Hadamard gate, while the following symbol is the CNOT gate with the first qubit as the control bit. The gauge symbols at the end indicate measurement.

6.5.1 Exercise: Superdense Coding

This approach dates back to the 1970s, but it wasn't published until 1992 [9].

A person, typically called Alice, wants to send a simple message consisting of two classical bits to her friend Bob. In other words, her message will be one of the four alternatives 00, 01, 10 or 11.[17] Actually, she may be be able to do so by sending *a single qubit* instead of two classical ones. Suppose their mutual friend Charlie does what we did in Exercise 6.4.7, namely he prepares a Bell state. It could, for instance, be Φ^+ in Eq. (6.34a). Charlie sends one of the entangled particles/qubits to Alice and then the other one to Bob. Alice may impose gates on her single qubit. Depending on the message she wants to convey, she imposes one of the three Pauli matrices. Or she does nothing – or, in other words, she imposes the identity operator gate, I_2. The scheme is illustrated in Fig. 6.10.

Next, she sends her particle to Bob. The particle pair, which is still entangled, is reunited physically with Bob, who imposes the same gates on it as Charlie did in the first place – however, in reverse order.

For each of Alice's four possible operations, I_2, σ_x, σ_y, or σ_z, what will be the outcome of measuring each of the two qubits for Bob?

This may look like going through a whole lot of trouble for the three friends just to be able to send a single qubit instead of two classical ones. However, the example does demonstrate that a quantum approach to information processing allows for shortcuts unavailable in a classical approach. At least that's something.

In the next example, Alice and Bob will exchange a lot more than just a single bit – both classical and quantum ones.

6.5.2 Exercise: Quantum Key Distribution

The scenario is the following: Alice wants to be able to pass on a message to Bob, a much longer one this time, with certainty of not having been eavesdropped. This can

[17] Alice and Bob may have agreed on assigning meaning to each of these alternatives beforehand – such as 'All is fine', 'I miss you', 'Please send money' and 'Come rescue me!', respectively.

actually be done. The scheme requires that they share *two* communication channels – a classical one, such as a phone line, and a quantum channel. Again, we will illustrate the protocol using spin-1/2 particles.

(a) Some purely classical considerations first: imagine that some message has been encoded as a long sequence of zeros and ones. Imagine also that Alice and Bob – and no one else – hold an encryption key, also consisting of zeros and ones, which is as long as the original message. Next, Alice adds the message and the key, modulo 2, bit by bit,[18] and broadcasts it – publicly. She does not have to worry about anyone intercepting the message, they will not be able to decipher it.

Why is that?

(b) We now set out to construct such an encryption key. Alice sends out spin-1/2 particles to Bob through a Stern–Gerlach apparatus which can be rotated 90 degrees between each emission, switching between x and z spin projection measurements.[19] For each particle Alice registers whether it is an x or a z eigenstate, and what eigenvalue it has – whether it has a positive or a negative spin projection.

For every particle that Bob receives from Alice, he picks a direction to orient his own apparatus – x or z – and measures the particle's spin projection along this axis. Sometimes he picks the basis in which the state is an eigenstate, sometimes he does not. He cannot tell in advance whether he has chosen the 'right' axis or not. Each and every time he will find the eigenvalue $+\hbar/2$ or $-\hbar/2$. He translates a positive measurement into '0' and a negative one into '1'.

Afterwards, with their classical communication channel, Alice and Bob may compare their choices of axes, bit by bit, and agree on a key – without actually sharing any content of the key itself.

How? Does it matter if their conversation is tapped?

What you learned in Exercise 4.3.2 may come in handy. And, perhaps, you will also find Table 6.1 useful. Here we have sketched a possible scenario for the first 12 spin-1/2 particles that Alice sends off. Which bits can be used for their key here? Where can Alice and Bob be sure to have the same bit value?

Table 6.1 This potential scenario demonstrates how Alice and Bob may agree on a secret key – provided that they haven't been eavesdropped. This visualization is inspired by Ref. [8].

Alice's state	$\leftarrow$	$\uparrow$	$\uparrow$	$\rightarrow$	$\downarrow$	$\rightarrow$	$\leftarrow$	$\downarrow$	$\downarrow$	$\uparrow$	$\downarrow$	$\leftarrow$	$\cdots$
Alice's bit	1	0	0	0	1	0	1	1	1	0	1	1	$\cdots$
Bob's choice of basis	X	X	Z	Z	X	Z	Z	Z	X	Z	X	X	$\cdots$
Bob's readoff	$\leftarrow$	$\rightarrow$	$\uparrow$	$\downarrow$	$\leftarrow$	$\downarrow$	$\downarrow$	$\downarrow$	$\leftarrow$	$\uparrow$	$\rightarrow$	$\leftarrow$	$\cdots$
Bob's bit	1	0	0	1	1	1	1	1	1	0	0	1	$\cdots$

[18] This is simple arithmetic: $0 + 1 = 1 + 0 = 1$ and $0 + 0 = 1 + 1 = 0$.

[19] It would be tempting to call the sender Walther instead of Alice. However, the name Alice is fixed by tradition.

Table 6.2 This potential scenario demonstrates how Alice and Bob may reveal that Eve has been trying to intercept their key. The visualization is inspired by Ref. [8].

Alice's state	←	↑	↑	→	↓	→	←	↓	↓	↑	↓	←	…
Alice's bit	1	0	0	0	1	0	1	1	1	0	1	1	…
Eve's choice of basis	Z	Z	X	Z	X	X	X	Z	Z	X	X	Z	…
Eve's readoff	↑	↑	←	↓	←	→	←	↓	↓	→	→	↓	…
Eve's bit	0	0	1	1	1	0	1	1	1	0	0	1	…
Bob's choice of basis	X	X	Z	Z	X	Z	Z	Z	X	Z	X	X	…
Bob's readoff	→	←	↑	↓	←	↓	↑	↓	←	↓	→	←	…
Bob's bit	0	1	0	1	1	1	0	1	1	1	0	1	…

(d) Eve, the eavesdropper,[20] wants to steal the key so that she can decipher the message later on. She may be able to intercept Alice and Bob's phone call where they discuss their choices of projection axes – x or z. But she knows that this will not be enough, she must also intercept the particles sent through the quantum channel, where Alice sends her string of carefully prepared spin-1/2 particle states. For each of these particles, Eve measures the spin projection somehow and then sends it off to Bob – hoping that he won't be able to tell that she has read it off.

However, if Eve actually has tried to pull off such a scam, Alice and Bob may, using their classical communication channel, easily reveal that Eve has been tampering with their signal. How?

Perhaps Table 6.2 could assist your train of thought here? In this scenario Alice has sent off exactly the same sequence of carefully prepared spin-1/2 particles as in Table 6.1. But this time Eve intercepts them and reads them off before passing them on to Bob. Just like Bob, she needs to choose an axis for which she measures the spin projection. Can you identify the specific bits where Eve reveals herself?

This key distribution scheme is often abbreviated as BB84 – after the inventors, Charles Bennett[21] and Gilles Brassard, and the year of its invention [8]. The beauty of it is that no eavesdropper can acquire the key without revealing herself. With access to the quantum channel she may perform her own measurements on the particles transmitted through. And by tapping the phone call she also knows which particles/qubits are used to construct the key and which ones are dropped by Alice and Bob. But only afterwards. She did *not* know which basis to use when she measured them. Just like Bob, Eve simply has to decide for herself if she wanted to measure the x-projection or the z-projection. Suppose Alice sent away $\chi_\leftarrow$ and Bob reports having performed an x-measurement for that particular particle. Then Eve knows that this particle counts, but she didn't know that – nor which basis to use – when she actually read it off. If

[20] Admittedly, this pun is quite corny. But tradition is to blame.

[21] The attentive reader may have noticed that Charles Bennett was also one of the people behind the superdense coding scheme.

she happened to measure the z-projection, she may have got the wrong result. Her real problem, however, is that she has collapsed the state from $\chi_\leftarrow$ into $\chi_\uparrow$ or $\chi_\downarrow$. So when Bob later on measures the spin projection in the x-basis, he may very well end up with a spin-right result, $\chi_\rightarrow$. This is exactly what happens for the first spin particle in Table 6.2.

If Alice and Bob now compare a relatively short part of their key over the phone, a large fraction of the bits, which should have coincided, will differ. And Eve is revealed.

6.6 Adiabatic Quantum Computing

Adiabatic quantum computing, and the closely related notion of *quantum annealing*, represents an alternative, or rather a supplement, to the gate-based approach to quantum computing addressed above.[22] Although the adiabatic and the gate-based approaches to quantum computing have been proven to be formally equivalent, it is fair to say that they are quite different – both conceptually and in terms of implementation. Adiabatic quantum computing schemes are generally limited to solving optimization problems. In this sense, it is more restricted than universal, gate-based quantum computing. However, efficient optimization problems are most certainly of interest when it comes to practical problem solving – for instance within logistics and machine learning, or, as we learned in Chapter 3, within quantum physics and chemistry itself for that matter. A quantum advantage in this regard would be most welcome. Adiabatic quantum computing also has the advantage of being more robust against noise and less prone to *decoherence* than gate-based quantum computing.

As the name suggest, adiabatic quantum computing exploits the adiabatic theorem, with which we became acquainted in Section 5.5. In such schemes, a quantum system starts out in the ground state of a well-known, easily implementable Hamiltonian H_I and then, slowly, the Hamiltonian is evolved into another one, H_F – one whose ground state somehow solves our minimization[23] problem. In effect, our system is subject to the time-dependent Hamiltonian

$$H(t) = s(t)H_I + (1 - s(t))H_F, \tag{6.36}$$

where both H_I and H_F are time independent and $s(t)$ slowly and continuously changes from 1 to 0. If we manage to encode the function we wish to minimize into the ground state of H_F somehow, we should be able find the solution as long as $s(t)$ evolves slowly enough for our system to stay in the ground state throughout the process. In the crossover between H_I and H_F, it is a clear advantage if the dynamic ground state evades avoided crossings – in accordance with the discussion following Exercise 5.5.1.

[22] We will not distinguish between adiabatic quantum computing and quantum annealing here; their distinction is, let's say, rather elusive.

[23] As optimization and minimization are two sides of the same coin, the processes only differ by the sign of the function in question; we hope that you will forgive us for using these terms interchangeably.

We will illustrate adiabatic quantum computing by a simple example which is strongly inspired by Ref. [17].

6.6.1 Exercise: Quantum Minimization

Suppose we set out to find the global minimum of this function:[24]

$$V_F(x) = (x^2 - 1)^2 - x/5. \tag{6.37}$$

We let this function be the potential of a single-particle Hamiltonian, Eq. (2.8), which we take to be our H_F. As the potential of our initial Hamiltonian, H_I, we take the harmonic oscillator, Eq. (3.9). Finally, let $s(t)$ evolve according to

$$s(t) = \frac{1}{2}\left(1 + \cos\left(\frac{\pi}{T}\,t\right)\right), \quad t \in [0, T]. \tag{6.38}$$

(a) Whereabouts do you expect the global minimum of $V_F(x)$ to be? Perhaps it is easy to estimate from looking at the functional form of Eq. (6.37). Or you could simply plot it – or have a look at the time development of the potential in Fig. 6.11. You could, of course, also do it the proper way by determining the roots of $V_F'(x)$.

(b) Solve the Schrödinger equation for this system with various choices for T and monitor the evolution as you go along. For our H_I, the ground state is known analytically. With m, $\hbar$ and k in Eq. (3.9) all being equal to unity, it reads

$$\psi_0(x) = \pi^{-1/4} e^{-x^2/2}. \tag{6.39}$$

Your implementation from Exercise 5.5.2 may be useful. As in that exercise, you could use the split-operator scheme, Eq. (5.25), with $\hat{A}$ being the kinetic energy and $\hat{B}$ the time-dependent potential. However, if you are a bit stingy when it comes to

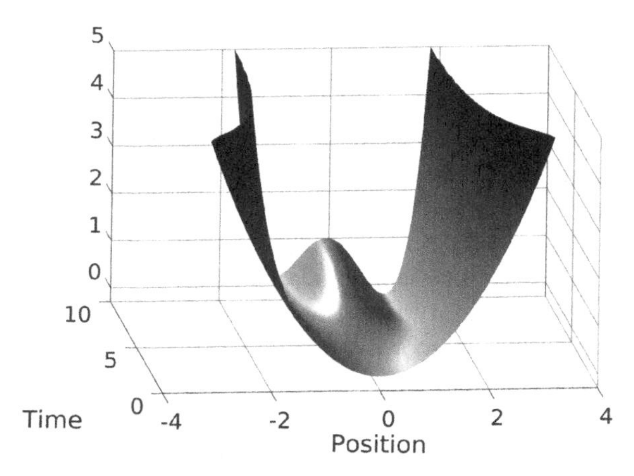

Figure 6.11 This surface illustrates the slowly varying potential of the Hamiltonian in Eq. (6.36), in which the harmonic oscillator potential, Eq. (3.9), evolves into Eq. (6.37).

[24] We do realize that this is not too hard in the first place. Just bear with us for a while, OK?

the numerics, you could also simply do a full matrix exponentiation at each time step, Eq. (5.23b), with $\overline{H} \approx H(t + \Delta t/2)$.

How large must the duration T be in order to ensure, more or less, adiabatic evolution?

(c) While the ground state of H_F encapsulates the minimum of Eq. (6.37), it is far from localized. And the ground state's preference for the global minimum to the right over the local one to the left could certainly be stronger. In order to remedy this, we will, quite artificially, start increasing the mass of the 'particle'. Thus, the time dependence shifts from the potential to the kinetic energy term. This is, in effect, a way to slowly and adiabatically turn off quantum effects and make the system increasingly classical.

From $t = T$ and onwards, our new time-dependent Hamiltonian is thus

$$H_{\rm loc}(t) = \frac{\hbar^2}{2m(t)}\frac{d^2}{dx^2} + V_F(x). \tag{6.40}$$

We will use this time-dependent 'mass':

$$m(t) = m_0(1 + 0.01(t - T)^2), \tag{6.41}$$

where $m_0 = 1$ is the initial mass.

Keep on evolving our system for $t > T$ with this Hamiltonian and see if we can, more or less, localize our wave function around the minimum near $x = 1$.

If you applied a split-operator technique in your implementation from (a), $\hat{A}$ and $\hat{B}$ should switch roles now. The middle of the sandwich, the time-dependent kinetic energy propagation, can be done efficiently by means of Fourier transforms as in Eq. (2.15). Or you could continue to calculate the full matrix exponential on the fly.

Hopefully, you agree that, in the end, a position measurement should be likely to yield a result very close to the minimum of our potential. Admittedly, it hardly seems worthwhile to go through all the trouble of solving the time-dependent Schrödinger equation, not to mention implementing it on an actual quantum computer, in order to find the minimum of a function as simple as the one in Eq. (6.37). The advantage of such approaches comes into play for more complex optimization problems, problems involving several variables and qubits.

We do hope that the example serves to illustrate how quantum behaviour may be exploited in order to avoid getting stuck in local minima. This is a severe problem for more conventional optimization schemes, which typically are based on some adaptation of the gradient descent method, which we first encountered in Exercise 3.4.4.

If you are familiar with *simulated annealing*, which is a classical approach also aiming to reduce the risk of getting stuck in local minima, it should not be too hard to see the motivation for the term *quantum* annealing. Quantum annealing has the advantage over simulated annealing that a quantum system may escape local minima by *tunnelling*.

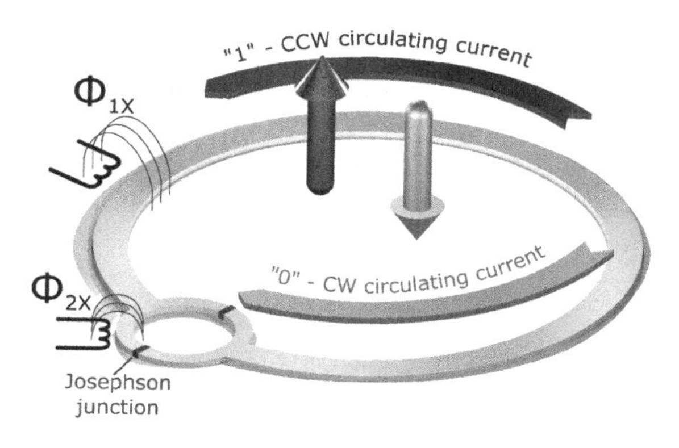

Figure 6.12 Schematic of the quantum bit used in the D-Wave One quantum annealer. In such a setup, two pieces of superconducting material are separated by a small barrier through which a current will pass without any external voltage. This current is quantized, and the two lowest current states, one corresponding to a current circulating clockwise and the other counter-clockwise, constitute the qubit. Media courtesy of D-Wave.

Perhaps placing Exercise 6.6.1 under the *technology* heading is stretching it a bit. However, the first commercially available quantum computer, the D-Wave One (see Fig. 6.12), was, in fact, a quantum annealer. Their quantum hardware implements *Ising models*. An Ising model may be seen as an arrangement of several spin-1/2 particles that all interact with their neighbours – a generalization of the two-spin system we saw in Exercise 5.1.4, in other words. The Hamiltonians of such Ising models are used to encode the function to be minimized – functions that take binary numbers as inputs and outputs.

7 Beyond the Schrödinger Equation

Thus far, we have referred to the Schrödinger equation as if it were the one, fundamental equation in quantum physics. It's not.

We will get acquainted with two rather different kinds of generalizations of the Schrödinger equation. First, we will address how quantum physics may be formulated in a manner consistent with relativity.[1] Second, we will discuss reversibility and openness in quantum physics.

7.1 Relativistic Quantum Physics

By now we have learned that things become rather strange when they become small – when quantum physics takes over and Newtonian mechanics no longer applies. However, Albert Einstein taught us that even large things look quite different from what we are used to when they reach velocities comparable with the speed of light (Fig. 7.1). According to his special theory of relativity, the kinetic energy of a classical, free particle of mass m is

$$T_{\text{rel}} = mc^2 \left(\sqrt{1 + \left(\frac{p}{mc} \right)^2} - 1 \right), \tag{7.1}$$

where c is the speed of light. Clearly, this differs from the expression we have been using: $T = p^2/2m$. The latter is only a decent approximation when we are dealing with objects moving a lot slower than light. Usually we are, but sometimes we are not.

7.1.1 Exercise: Relativistic Kinetic Energy

Make a plot of T_{rel} along with the kinetic energy expression we have used thus far, $T = p^2/2m$, as functions of the momentum p.

Prove that $T_{\text{rel}} \approx p^2/2m$ when $(p/mc)^2 \ll 1$.

You could set the mass m to 1 kg and use SI units. However, it may be more convenient to let mc^2 be your energy unit and introduce the variable $x = (p/mc)^2$. For the sake of clarity: this exercise has nothing to do with quantum physics as such.

[1] *Special* relativity, that is.

Figure 7.1 Like quantum physics, special relativity has some consequence that are quite far from intuitive. This cartoon illustrates the phenomenon of *time dilation*; time progresses more slowly for an observer moving at a velocity close to the speed of light than for one at rest.

There are situations in which quantum particles are accelerated towards relativistic speeds – velocities that constitute a significant fraction of the speed of light. For instance, with state of the art lasers, electrons may be driven towards speeds for which a non-relativistic treatment is inadequate. And relativity certainly plays a role in high-energy physics facilities, such as CERN, where particles are collided with other particles to create new ones. For atoms and molecules at rest with large, highly charged nuclei, the innermost electrons may be accelerated towards the speed of light by the Coulomb attraction from the nucleus. In fact, this is the reason why gold looks like gold and not silver.

In other words, in both dynamical and stationary situations, quantum equations consistent with special relativity are sometimes needed. A natural starting point in this regard would be the following energy relation from special relativity:[2]

$$E^2 = m^2c^4 + p^2c^2. \tag{7.2}$$

The (time-dependent) Schrödinger equation comes about by insisting that the action of the energy operator, the Hamiltonian, on the wave function, $\hat{H}\Psi$, provides the same result as the action of $i\hbar\partial/\partial t$ on Ψ. If we follow this train of thought and turn the above energy expression into an operator,

$$\hat{H}^2 = m^2c^4 + \hat{p}^2c^2, \tag{7.3}$$

we arrive at the Klein–Gordon equation for a free particle:

$$-\hbar^2\frac{\partial^2}{\partial t^2}\Psi = \left[-\hbar^2c^2\nabla^2 + m^2c^4\right]\Psi. \tag{7.4}$$

[2] In the special case of a particle at rest, $p = 0$, a much more famous special case of Eq. (7.2) emerges.

With a potential it reads[3]

$$\left[i\hbar\frac{\partial}{\partial t} - V(\mathbf{r})\right]^2 \Psi = \left[-\hbar^2 c^2 \nabla^2 + m^2 c^4\right] \Psi. \tag{7.5}$$

Actually, it was neither the Swede Oskar Klein nor the German Walter Gordon who arrived first at the equation that came to bear their names. It was actually Erwin Schrödinger himself; it was his first attempt at formulating a wave equation for matter. He dismissed it on the grounds that it did not provide the bound state energies for the hydrogen atom with sufficient precision. Instead he settled on a non-relativistic formulation. At the time, he could hardly have been expected to know why his first attempt 'failed;' the reason is not intuitive at all.

This issue was eventually resolved after the Englishman Paul Dirac, in 1928, had proposed an alternative relativistic equation. Dirac, who you can see in the middle of the crowd in Fig. 1.3, didn't appreciate the fact that the Klein–Gordon equation is second order in both time and space. Instead he insisted on a linear equation:[4]

$$i\hbar\frac{\partial}{\partial t}\Psi = \left[c\boldsymbol{\alpha} \cdot \hat{\mathbf{p}} + mc^2\beta\right] \Psi, \tag{7.6}$$

with $\boldsymbol{\alpha} = [\alpha_x, \alpha_y, \alpha_z]$ in the three-dimensional case. In retrospect, it seems fair to say that it comes off as rather surprising that what appears to be an overly pragmatic approach would actually lead to such a huge breakthrough.

7.1.2 Exercise: Arriving at the Dirac Equation

Dirac demanded that his Hamiltonian,

$$\hat{H} = c\boldsymbol{\alpha} \cdot \hat{\mathbf{p}} + mc^2\beta, \tag{7.7}$$

should fulfil Eq. (7.3). This may appear rather unreasonable. And, indeed, it cannot be achieved if the components of $\boldsymbol{\alpha}$ and β are numbers. It can, however, if they are *matrices*.

(a) In the one-dimensional case, where we only have α_x, no α_y nor any α_z, show that Eq. (7.3) leads us to require the following:

$$\alpha_x^2 = \beta^2 = I, \quad \alpha_x\beta + \beta\alpha_x = 0, \tag{7.8}$$

where I is the identity matrix and '0' here is to be understood as the zero matrix.

Also, show by direct calculation that these equations are satisfied with these choices:

$$\alpha_x = \sigma_x = \begin{pmatrix} 0 & 1 \\ 1 & 0 \end{pmatrix}, \quad \beta = \sigma_z = \begin{pmatrix} 1 & 0 \\ 0 & -1 \end{pmatrix}. \tag{7.9}$$

[3] For the record: we have not really explained how the potential energy actually enters into the equation here.

[4] This version corresponds to a time-independent Hamiltonian without any potential.

(b) In the three-dimensional case, explain why these matrices must fulfil the following relations:

$$\alpha_k \alpha_l + \alpha_l \alpha_k = 0 \quad \text{for} \quad k \neq l, \tag{7.10a}$$

$$\alpha_k \beta + \beta \alpha_k = 0, \tag{7.10b}$$

$$\alpha_k^2 = I, \tag{7.10c}$$

$$\beta^2 = I, \tag{7.10d}$$

where k, l are either x, y or z. In other words, all of these matrices must be their own inverses, and they must all anti-commute, Eq. (4.18), with each other.

In order to fulfil all of these equations, the matrices cannot be any smaller than 4×4 matrices.

(c) Verify that these choices indeed fulfil Eqs. (7.10):

$$\alpha_x = \begin{pmatrix} 0_{2\times 2} & \sigma_x \\ \sigma_x & 0_{2\times 2} \end{pmatrix} = \begin{pmatrix} 0 & 0 & 0 & 1 \\ 0 & 0 & 1 & 0 \\ 0 & 1 & 0 & 0 \\ 1 & 0 & 0 & 0 \end{pmatrix},$$

$$\alpha_y = \begin{pmatrix} 0_{2\times 2} & \sigma_y \\ \sigma_y & 0_{2\times 2} \end{pmatrix} = \begin{pmatrix} 0 & 0 & 0 & -\mathrm{i} \\ 0 & 0 & \mathrm{i} & 0 \\ 0 & -\mathrm{i} & 0 & 0 \\ \mathrm{i} & 0 & 0 & 0 \end{pmatrix},$$

$$\alpha_z = \begin{pmatrix} 0_{2\times 2} & \sigma_z \\ \sigma_z & 0_{2\times 2} \end{pmatrix} = \begin{pmatrix} 0 & 0 & 1 & 0 \\ 0 & 0 & 0 & -1 \\ 1 & 0 & 0 & 0 \\ 0 & -1 & 0 & 0 \end{pmatrix},$$

$$\beta = \begin{pmatrix} I_2 & 0_{2\times 2} \\ 0_{2\times 2} & -I_2 \end{pmatrix} = \begin{pmatrix} 1 & 0 & 0 & 0 \\ 0 & 1 & 0 & 0 \\ 0 & 0 & -1 & 0 \\ 0 & 0 & 0 & -1 \end{pmatrix}.$$

The 2×2 off-diagonal blocks of the α-matrices are the Pauli matrices, which we should know rather well by now.

Equation (7.6), where the matrices comply with Eqs. (7.10), is the Dirac equation for a free particle.[5] If we introduce a potential $V(\mathbf{r})$ and an external electromagnetic field given by the vector potential $\mathbf{A}(\mathbf{r}, t)$, it reads

$$\mathrm{i}\hbar \frac{\partial}{\partial t} \Psi(\mathbf{r}, t) = \left[c\boldsymbol{\alpha} \cdot \left(\hat{\mathbf{p}} - q\mathbf{A}(\mathbf{r}, t) \right) + V(\mathbf{r}) + mc^2 \beta \right] \Psi(\mathbf{r}, t). \tag{7.11}$$

As we have seen, the Hamiltonian is a 4×4 matrix in the three-dimensional case. Note that the Hamiltonian is not a matrix in the same sense as we saw in Section 2.2;

[5] In the literature you will frequently see the Dirac equation formulated in a more compact fashion. While simplifying the notation, this hardly simplifies actually solving it.

it is a matrix acting on a wave function with four components, which, in turn, often is written as a vector with two *spinors*:

$$\Psi(\mathbf{r};t) = \begin{pmatrix} \Psi_1(\mathbf{r};t) \\ \Psi_2(\mathbf{r};t) \\ \Psi_3(\mathbf{r};t) \\ \Psi_4(\mathbf{r};t) \end{pmatrix} = \begin{pmatrix} \mathbf{\Phi}_+(\mathbf{r};t) \\ \mathbf{\Phi}_-(\mathbf{r};t) \end{pmatrix}. \tag{7.12}$$

Here $\mathbf{\Phi}_+$ and $\mathbf{\Phi}_-$ have two components each. It came to be realized that these components are related to *spin*; the first component corresponds to the spin of the particle pointing upwards and the second one corresponds to spin downwards:

$$\mathbf{\Phi}_+ = \begin{pmatrix} \Psi_\uparrow^+(\mathbf{r};t) \\ \Psi_\downarrow^+(\mathbf{r};t) \end{pmatrix}. \tag{7.13}$$

Note that this form is nothing new; we saw it in Eqs. (4.3) and (6.20):

$$\mathbf{\Phi}_+(\mathbf{r};t) = \Psi_\uparrow^+(\mathbf{r};t)\begin{pmatrix} 1 \\ 0 \end{pmatrix} + \Psi_\downarrow^+(\mathbf{r};t)\begin{pmatrix} 0 \\ 1 \end{pmatrix} = \Psi_\uparrow^+\,\chi_\uparrow + \Psi_\downarrow^+\,\chi_\downarrow. \tag{7.14}$$

With this realization, we are touching upon what Schrödinger couldn't have known back in the 1920s: the Dirac equation describes a particle with spin one-half – which has the two possible spin projections $1/2\,\hbar$ (up) and $-1/2\,\hbar$ (down). In other words, for an electron, the Dirac equation is in fact the proper dynamical equation. The Klein–Gordon equation, as it turns out, happens to describe spinless particles. This is the reason why it produces (slightly) erroneous energies for the electron in a hydrogen atom. The solutions for the time-independent Dirac equation, on the other hand, provide eigenenergies with small deviations from the solutions of the Schrödinger equation, Eq. (3.4) – deviations that actually agree with experiments.

As in the case of the Schrödinger equation, the time-independent versions of Eqs. (7.5) and (7.11), without any time-dependent vector potential $\mathbf{A}$ in the latter case, are obtained by substituting $\mathrm{i}\hbar\partial/\partial t$ with the eigenenergy E.

But what about the spinors of Eq. (7.12); why are there two of them? What's the meaning of the lower component $\mathbf{\Phi}_-$? Well, just like the emergence of spin, this lower component brought about significant new physics which nobody expected at that time. We will investigate this physics a bit further by solving the time-independent Dirac equation.

7.1.3 Exercise: Eigenstates of the Dirac Hamiltonian

In Chapter 3 we studied the energy spectra of bound states by solving the time-independent Schrödinger equation, Eq. (3.1), in one dimension. We will now do the same for the time-independent one-dimensional Dirac equation:

$$\left[-\mathrm{i}\hbar c\sigma_x \frac{\mathrm{d}}{\mathrm{d}x} + V(x)I_2 + mc^2\sigma_z\right]\psi(x) = \varepsilon\,\psi(x), \tag{7.15}$$

where we have, in accordance with Eq. (7.9), set $\alpha_x = \sigma_x$ and $\beta = \sigma_z$. In this case, the wave function has two (scalar) components:

$$\psi(x) = \begin{pmatrix} \psi_+(x) \\ \psi_-(x) \end{pmatrix}, \tag{7.16}$$

and the Hamiltonian is a 2×2 matrix:

$$H = \begin{pmatrix} V(x) + mc^2 & -i\hbar c \frac{d}{dx} \\ -i\hbar c \frac{d}{dx} & V(x) - mc^2 \end{pmatrix}. \tag{7.17}$$

In order to compare your relativistic spectrum with the non-relativistic one, take your implementation from Exercise 3.1.2 as your starting point. Let your potential be given by Eq. (2.30) with width $w = 1$ and smoothness $s = 10$. We will be adjusting the depth $-V_0$. Set the speed of light c to 137, in addition to $m = 1$ and $\hbar = 1$.[6]

As usual, discretize our spatial domain into $n+1$ grid points distributed over a finite extension L; it does not have to be very large, 10 or so would suffice.

The Hamiltonian of Eq. (7.20) now becomes a $2n \times 2n$ matrix, which makes it computationally more expensive to diagonalize than its non-relativistic counterpart, which is an $n \times n$ matrix. It may still be done within reasonable precision, however, using standard numerical tools. Along the diagonal of this matrix you will have $V + mc^2 I_n$ along the first half and $V - mc^2 I_n$ along the second. The off-diagonal block consists of the momentum operator. Both this and the kinetic energy operator in the non-relativistic case should be implemented using FFT.[7]

Once you have expanded your implementation from Exercise 3.1.2, try to answer the following questions:

(a) Start with a comparatively shallow well, $V_0 = -10$ or so, and check out the relativistic eigenenergies. You can simply plot the sorted eigen-energies in ascending order against their indices. The relativistic spectrum has one pronounced trait very different from the non-relativistic one, doesn't it?

(b) Next, consider only bound states. For the Dirac Hamiltonian this would mean the energies that are positive, but less than the mass energy mc^2. Compare these energies, with the mass energy mc^2 subtracted, to the non-relativistic ones. With a potential that is not very deep, they should coincide.

Do they?

(c) Now, start lowering your well. As you do so, relativistic and non-relativistic energies will start to differ. Do so and compare repeatedly as you increase $|V_0|$. How deep must the well be before you start seeing deviations?[8] And where do the spectra differ? You may have to make your well quite deep before you see anything.

---

[6] The number 137 may appear somewhat far fetched. It's not, however; it is the speed of light in atomic units.

[7] As may become apparent after having done Exercise 7.1.4, your implementation should be such that the square of the momentum operator actually coincides, to a reasonable degree, with the operator for the momentum squared. Mathematically, it is always so, but it certainly doesn't have to be numerically.

[8] Do make sure that any discrepancies you may see are not due to insufficient numerical precision.

The first striking observation is the fact that the Dirac spectrum separates into two very distinct parts, one below $-mc^2$ and one close to or above mc^2. Because the mass energy term, βmc^2, is so large, this may have been obvious from the outset. What is not at all obvious, however, is how to *interpret* these negative energy solutions. To make a long story short, they relate to *anti-particles*. As it turns out, every type of particle has its own anti-particle with the same mass and spin, but opposite charge.[9] While it took some time to realize, the existence of anti-matter is well established now. It is also realized that pairs of particles and anti-particles may be created and annihilated; they may appear and disappear. One consequence of this, in turn, is that just solving the Dirac equation is not sufficient in order to describe relativistic quantum processes in the high-energy regime. For this you need a *field theoretical description*, in which the number of particles is not fixed. We will not venture into that topic here.

As for the relativistic corrections to the spectrum of bound states, you may have found that the discrepancies are larger for the higher energies – closer to the threshold at zero. One simple way of acquiring some intuition for this is provided by the notion of *relativistic mass*. Sometimes it makes sense to consider a relativistic particle by means of increased inertia. The particle is subject to an effective increase in the mass m when the kinetic energy is high, as would be the situation for the highest energies within our deep confining potential. Such an increase in mass will, in turn, effectively reduce the kinetic energy.

In Exercise 7.1.1 we saw that in the limit that the speed of the particle is a lot less than the speed of light, the expression for kinetic energy according to special relativity coincides with the 'old' Newtonian expression. Shouldn't the same be the case within quantum physics? Or, in other words, shouldn't both the Klein–Gordon and the Dirac equations agree with the Schrödinger equation as long as the probability for the particle to attain relativistic momenta is negligible? After all, this is what we saw above; with moderate depth $-V_0$ in Exercise 7.1.3, there wasn't really any difference between the relativistic and the non-relativistic spectrum – besides the negative energy part, that is.

For the Klein–Gordon equation it is more or less straightforward to see that it reproduces the Schrödinger equation at low kinetic energy as the Hamiltonian is given directly by Eq. (7.2):

$$E = \sqrt{m^2c^4 + p^2c^2} = mc^2 + \frac{p^2}{2m} - \mathcal{O}\left(\frac{p^4}{m^2c^4}\right). \tag{7.18}$$

The Dirac equation, Eq. (7.11), on the other hand, doesn't look anything like the Schrödinger equation, which for a single particle in an electromagnetic field reads

$$i\hbar\frac{\partial}{\partial t}\Psi = \left[\frac{1}{2m}\left(\hat{\mathbf{p}} - q\mathbf{A}\right)^2 + V(\mathbf{r})\right]\Psi. \tag{7.19}$$

Well, let's have a closer look.

[9] Actually, particles have several kinds of charge, not just electric ones – all of which are opposite for particles and anti-particles.

7.1.4 Exercise: The Non-relativistic Limit of the Dirac Equation

Again, we start out with the one-dimensional time-dependent Dirac equation with the Hamiltonian of Eq. (7.17). Actually, it is more convenient to shift the energy downwards[10] by the mass energy mc^2 so that our actual Hamiltonian will be

$$H = \begin{pmatrix} V(x) & -i\hbar c\frac{\mathrm{d}}{\mathrm{d}x} \\ -i\hbar c\frac{\mathrm{d}}{\mathrm{d}x} & V(x) - 2mc^2 \end{pmatrix}. \tag{7.20}$$

(a) Write the coupled equations in terms of $\Psi_+(x;t)$ and $\Psi_-(x;t)$ separately.

(b) In the non-relativistic limit it is reasonable to assume that both the time variation of Ψ_- and the influence of the potential V on Ψ_- is negligible compared to the large mass energy term. Explain how this leads to

$$i\hbar c\frac{\mathrm{d}}{\mathrm{d}x}\Psi_+ \approx -2mc^2\Psi_-. \tag{7.21}$$

(c) Now, if you solve this equation, algebraically, for Ψ_- and insert it into the equation for $i\hbar\partial\Psi_+/\partial t$, what do you get?

Hopefully, your answer to the last question was

$$i\hbar\frac{\partial}{\partial t}\Psi_+ = \left[-\frac{\hbar^2}{2m}\frac{d^2}{dx^2} + V(x)\right]\Psi_+ = \left[\frac{\hat{p}^2}{2m} + V(x)\right]\Psi_+, \tag{7.22}$$

which should appear quite familiar by now – despite the '+' subscript.

If we had exposed Eq. (7.11) to the same treatment,[11] we would have got something very similar to Eq. (7.19):

$$i\hbar\frac{\partial}{\partial t}\boldsymbol{\Psi}_+ = \left[\frac{1}{2m}(\hat{\mathbf{p}} - q\mathbf{A})^2 + V(x) - \frac{q\hbar}{2m}\mathbf{B}\cdot\boldsymbol{\sigma}\right]\boldsymbol{\Psi}_+. \tag{7.23}$$

There are two differences, though:

(1) there is an additional term in the Hamiltonian,

and

(2) the wave function now has two components, Eq. (7.13).

With $\hat{\mathbf{s}} = \hbar/2\,\boldsymbol{\sigma}$, we may recognize the extra term, the interaction between spin and a magnetic field, as Eq. (4.15). This is where the somewhat mysterious factor $g = 2$ comes from; a classical magnetic dipole would have $g = 1$.

Truth be told, there are a couple of terms missing from Eq. (7.23) – corrections to the interaction with the potential $V(x)$, which a more thorough treatment would have picked up.

[10] Such a shift is always admissible; it corresponds to a redefinition of the zero level for energy. In more technical terms, it shifts the wave function only by a global phase factor which does not alter any physics.

[11] Do not be afraid to try! You will need what you learned in Exercise 4.3.1 along with Eq. (2.6).

Note that there is an important difference between the one-dimensional and the three-dimensional situation here – in addition to the obvious: there simply isn't any *spin* in the one-dimensional case. Thus, you may rightfully question whether it made sense to include spin in the one-dimensional examples we looked at in Chapter 4. In our defence, the lessons learned in that chapter do carry over to the more realistic three-dimensional case. Moreover, a real system with three spatial dimensions could be confined in two of them so that it, effectively, becomes one-dimensional.

And now for something completely different.

7.2 Open Quantum Systems and Master Equations

Apart from a brief outline at the beginning of Section 3.4, we have thus far assumed that our quantum system is unaffected by the world around it – except when we perform measurements on it. This is not always a reasonable assumption. Experiments in quantum physics are often conducted in (almost) vacuum, at low temperatures, with very dilute gases and within apparatus which tries to shield off the omnipresent background radiation. However, despite such efforts to minimize the influence of the environment, sometimes you simply have to include the interaction between your quantum system and its surroundings in order to make sensible predictions.

This is bad news. While the quantum system you want to describe theoretically or computationally may be complicated enough, this complexity cannot hold a candle to the complexity involved in describing its surroundings. In most cases it has so many degrees of freedom that you simply cannot include them in your Schrödinger equation if you have any ambition of actually solving it. So we must settle for trying to include the interactions with the surroundings in a simplified, effective manner somehow. In doing so, we typically end up replacing our original wave function equation, usually the Schrödinger equation, with a *master equation*.

Before we can address the notion of master equations, we need to say a word or two about *density matrices*. For a quantum system in a state given by the wave function Ψ, the density matrix, which we call $\hat{\rho}$, is an operator that simply extracts the projection of another state Φ onto Ψ. Dirac notation[12] is particularly convenient for expressing such operators:

$$\hat{\rho} = |\Psi\rangle\langle\Psi|. \tag{7.24}$$

Although $\hat{\rho}$ strictly speaking is an operator, not necessarily a matrix, we see from the above expression that if $|\Psi\rangle$ is a column vector, $\langle\Psi| = |\Psi\rangle^\dagger$ is a row vector and $\hat{\rho}$ is, indeed, a matrix.

[12] Yes, it is the same Dirac as above.

7.2.1 Exercise: The von Neumann Equation

With the state $|\Psi\rangle$, which now has acquired brackets, following the Schrödinger equation, $i\hbar\partial/\partial t\,|\Psi\rangle = \hat{H}|\Psi\rangle$, explain why $\hat{\rho}$ in Eq. (7.24) follows the equation

$$i\hbar\frac{\partial}{\partial t}\hat{\rho} = \left[\hat{H}, \hat{\rho}\right]. \tag{7.25}$$

Here $[\hat{A}, \hat{B}] = \hat{A}\hat{B} - \hat{B}\hat{A}$ is the *commutator*, which we first encountered in Exercise 2.6.2.

Equation 7.25 is called the *von Neumann equation*. For a density matrix of the form of Eq. (7.24), a so-called *pure state*, it is entirely equivalent to the Schrödinger equation. 'So what's the point?', you may rightfully ask. So far, introducing density matrices has only led to a rephrasing of what we already know – in a slightly more cumbersome way.

Our motivation was, however, to include the influence of the environment somehow. This may be achieved by introducing *partial traces*.

As mentioned in connection with Exercise 6.4.6(d), the trace of a matrix is simply the sum of its diagonal elements. The trace of a more general operator $\hat{A}$ is defined as

$$\mathrm{Tr}\,\hat{A} = \sum_n \langle \alpha_n|\hat{A}|\alpha_n\rangle, \tag{7.26}$$

where $|\alpha_n\rangle$ is an orthonormal basis for the full space in question. Any orthonormal basis will lead to the same trace.

Suppose now that a quantum system is made up of two parts. It could consist of two particles, such as the situations in Exercises 4.4.3 and 5.1.4. The two parts could consist of a small quantum system embedded in a larger system – the situation we addressed initially. Or it could be the more familiar case of a single particle with both spatial and spin degrees of freedom, as in Eqs. (4.3) and (4.6). If we want to disregard one of the sub-systems and only describe the other one, we could do so by *tracing out* all the degrees of freedom of the system we want to disregard. Let's take the state of Eq. (4.3), which, in general, is an entangled state of spatial and spin degrees of freedom, as an example. The full density matrix of this state is

$$\begin{aligned}\hat{\rho} = |\Phi\rangle\langle\Phi| &= \left(|\Psi_\uparrow\rangle|\chi_\uparrow\rangle + |\Psi_\downarrow\rangle|\chi_\downarrow\rangle\right)\left(|\Psi_\uparrow\rangle|\chi_\uparrow\rangle + |\Psi_\downarrow\rangle|\chi_\downarrow\rangle\right)^\dagger \\ &= |\Psi_\uparrow\rangle\langle\Psi_\uparrow|\;|\chi_\uparrow\rangle\langle\chi_\uparrow| + |\Psi_\uparrow\rangle\langle\Psi_\downarrow|\;|\chi_\uparrow\rangle\langle\chi_\downarrow| \\ &\quad + |\Psi_\downarrow\rangle\langle\Psi_\uparrow|\;|\chi_\downarrow\rangle\langle\chi_\uparrow| + |\Psi_\downarrow\rangle\langle\Psi_\downarrow|\;|\chi_\downarrow\rangle\langle\chi_\downarrow|,\end{aligned} \tag{7.27}$$

where $|\Psi_{\uparrow,\downarrow}\rangle$ and the spinors $|\chi_{\uparrow,\downarrow}\rangle$ reside in different mathematical spaces – position space and spin space, respectively. Correspondingly, '$\langle\Psi_\uparrow|\;|\chi_\uparrow\rangle$' and similar expressions in the above equation are not to be taken as any inner product.

Suppose now that we want to disregard the spin degree of freedom. Correspondingly, we *trace out* this degree of freedom – this means projecting from both left and right onto the spinors $\chi_\uparrow$ and $\chi_\downarrow$, respectively, and summing:

$$\hat{\rho} \to \hat{\rho}_{\mathrm{position}} = \langle\chi_\uparrow|\hat{\rho}|\chi_\uparrow\rangle + \langle\chi_\downarrow|\hat{\rho}|\chi_\downarrow\rangle. \tag{7.28}$$

This renders a much simpler *reduced* density matrix. The spin projections we imposed do not affect the spatial parts, the Ψ-parts of Eq. (7.27). Some of the inner products in spin space, on the other hand, end up as zero as dictated by Eqs. (4.5). With these equations we may find that the density matrix, after having traced out spin, reduces to

$$\hat{\rho}_{\text{position}} = |\Psi_\uparrow\rangle\langle\Psi_\uparrow| + |\Psi_\downarrow\rangle\langle\Psi_\downarrow|. \tag{7.29}$$

In the general case of a *bipartite system*, a system that resides in two mathematical spaces – let's call them space A and space B – the reduced density matrix corresponding to A is

$$\hat{\rho}_A = \mathrm{Tr}_B\, \hat{\rho} = \sum_n \langle \beta_n|\hat{\rho}|\beta_n\rangle, \tag{7.30}$$

where $\{|\beta_n\rangle\}$ is an orthonormal basis for the B space. In the above case, it consisted of $|\chi_\uparrow\rangle$ and $|\chi_\downarrow\rangle$.[13]

Equation (7.26) differs from Eq. (7.30) in that we only project onto some, not all, degrees of freedom for the bipartite system.[14]

While the full density matrix, Eq. (7.24), is a pure state, a *reduced* density matrix, Eq. (7.30), is generally not; it is usually *mixed*. It does not correspond to a single wave function, but rather to a set of several wave functions. We will return to this issue.

7.2.2 Exercise: Pure States, Entanglement and Purity

(a) Check that you actually arrive at Eq. (7.29) from Eq. (7.28) by imposing the projections onto the spin states in Eq. (7.27).
(b) When your wave function has the product form of Eq. (4.6), what is $\hat{\rho}_{\text{position}}$ then? Is it a pure state?
(c) To what extent a reduced density matrix differs from a pure state can be quantified. The *purity* is one way of doing so.[15]

$$\gamma = \mathrm{Tr}\,\hat{\rho}^2. \tag{7.31}$$

The purity γ is 1 for pure states only, and it takes a value between 0 and 1 for any mixed state. The lower γ is, the less pure – and more mixed – is the state.

Now, suppose some bipartite quantum system consisting of two qubits, A and B, is in the normalized state:

$$|\Psi\rangle = c_1|0\rangle_A|0\rangle_B + c_2|1\rangle_A|1\rangle_B = c_1|00\rangle + c_2|11\rangle, \tag{7.32}$$

where in the last equality we have adopted the somewhat more compact notation we used in Chapter 6. As long as none of the coefficients c_1 and c_2 are zero, this would correspond to an entangled state; for $c_1 = c_2 = 1/\sqrt{2}$ it is the Bell state of Eq. (6.34a) and for $c_1 = -c_2$ it is the Bell state of Eq. (6.34b).

[13] For simplicity, we have assumed that the B space has a countable basis. In the continuous case, the sum in Eq. (7.30) must be replaced with an integral.

[14] Formally, Eq. (7.30) would read $\mathrm{Tr}_B\,\hat{\rho} = \sum_n \left(\hat{I}_A \otimes \langle\beta_n|\right)\,\hat{\rho}\,\left(\hat{I}_A \otimes |\beta_n\rangle\right)$ in a more proper formulation, where $\hat{I}_A$ is the identity operator in the A-space.

[15] The von Neumann entropy, $S = -\mathrm{Tr}\,(\hat{\rho}\ln\hat{\rho})$, is another.

Show that

$$\hat{\rho}_A = |c_1|^2|0\rangle\langle 0| + \left(1 - |c_1|^2\right)|1\rangle\langle 1|, \tag{7.33}$$

where we have dropped the subscript indicating the space on the bras and kets.

(d) Plot the purity of $\hat{\rho}_A$ as a function of $|c_1|^2$. When is it minimal?

Hopefully, you found that the purity is minimal when the entanglement is maximal – for $|c_1|^2 = |c_2|^2$, or, equivalently, $c_2 = e^{i\phi}c_1$. Note also that the reduced density matrix of Eq. (7.33) is not able to include the information contained in the phase difference ϕ. Specifically, if the initial pure state, Eq. (7.32), is one of the Bell states of Eqs. (6.34), you cannot tell which one from $\hat{\rho}_A$.

7.2.3 Exercise: Two Spin-1/2 Particles Again

(a) Redo Exercise 5.1.4(a), except this time we will start out with the $\chi_\uparrow^{(1)}\chi_\downarrow^{(2)}$ state and plot the time-dependent probability of spin up for the first particle, with no regard for the spin of the second particle.

(b) Similar to what we did in Exercise 7.2.2(c) and (d), let your A-space be the spin of the first particle and B-space be the spin of the other particle. If you write down the reduced density matrix for particle 1, $\hat{\rho}_A$, in terms of the coefficients a, b, c and d in Eq. (5.20), you should find that

$$\hat{\rho}_A = \begin{pmatrix} |a|^2 + |b|^2 & ac^* + bd^* \\ a^*c + b^*d & |c|^2 + |d|^2 \end{pmatrix}. \tag{7.34}$$

Check that you actually arrive at Eq. (7.34).

Also, show that the probability you determined in (a) is the same as $\langle\chi_\uparrow|\hat{\rho}_A|\chi_\uparrow\rangle$, the first diagonal element of $\hat{\rho}_A$.

(c) The expectation value of some physical quantity Q with operator $\hat{Q}$ may be determined from a density matrix by the trace of their product:

$$\langle Q\rangle = \mathrm{Tr}\left(\hat{Q}\hat{\rho}\right). \tag{7.35}$$

For a pure state, this is consistent with our first definition, Eq. (1.25).

For the simulation you just performed, plot the expectation value of the spin projection along each of the three axes for particle A:

$$\langle s_x\rangle = \mathrm{Tr}\,(\hat{s}_x\,\hat{\rho}_A) = \frac{\hbar}{2}\mathrm{Tr}\,(\sigma_x\,\hat{\rho}_A), \tag{7.36}$$

and so on.

(d) Plot the absolute value of the off-diagonal elements of $\hat{\rho}_A$ as a function of time.

(e) Plot the purity of $\hat{\rho}_A$, $\gamma(t) = \mathrm{Tr}\,\hat{\rho}_A^2$, as a function of time.

The two-particle system starts out as a product state, and, correspondingly, the reduced density matrix $\hat{\rho}_A$ is pure at $t = 0$. During the interaction with the magnetic field, does it seem to become a pure state again at any point?

In the above examples we constructed reduced density matrices via the full wave functions. Although these examples may have shed some light on what reduced density matrices are, applying density matrices in situations where the full wave function is accessible usually wouldn't make too much sense.

It makes more sense to apply a description in terms of a reduced density matrix when the B system simply is too complicated to handle. This would be the situation if a comparatively small quantum system A interacts with some large environment B. Any ambition of describing the full wave function of such a bipartite system is usually only a pipe dream.

If we start out in a general product state, $|\Psi\rangle = |\Psi_a\rangle_A|\Psi_b\rangle_B$, the full wave function will follow the Schrödinger equation, but not the A and the B parts separately. As long as there is some kind of interaction between A and B, the full wave function quickly becomes an entangled one. Correspondingly, $\hat{\rho}_A(t)$ goes from being pure to being mixed. Being unable to resolve the evolution of the full wave function $|\Psi(t)\rangle$, we could try to settle for the next best thing – to describe the evolution of just $\hat{\rho}_A(t)$.

As a starting point, we would set up the von Neumann equation, Eq. (7.25), and trace out B. This would leave a messy equation which is not even local in time; it would depend on the *history* of the evolution, not just the present state. In order to arrive at something that can be solved in a meaningful way, several approximation are typically imposed – depending on the nature of the systems and their interaction. In certain situations we may reasonably impose assumptions about the B-system, such as assuming that it is in a so-called *thermal state*.

At the end of the day we hope to arrive at an equation for $\hat{\rho}_A$ that is local in time and only depends on the degrees of freedom pertaining to A. Even with no explicit reference to the degrees of freedom of the B-part, the interaction between A and B may still be incorporated in some effective manner. Such an equation governing the evolution of a density matrix is called a *master equation*.

When deriving master equations from more fundamental principles, we would typically hope to arrive at an equation consistent with this generic form:

$$i\hbar\frac{\partial}{\partial t}\hat{\rho} = \left[\hat{H}, \hat{\rho}\right] - \frac{i}{2}\sum_{kl}\Gamma_{k,l}\left(\hat{a}_k^\dagger\hat{a}_l\,\hat{\rho} + \hat{\rho}\,\hat{a}_l^\dagger\hat{a}_k - 2\hat{a}_l\,\hat{\rho}\,\hat{a}_k^\dagger\right). \tag{7.37}$$

If we remove the last terms on the right hand side, we recognize the von Neumann equation, Eq. (7.25), which, in turn, is equivalent to the Schrödinger equation in the case of a pure state. So Eq. (7.37) is a generalization of the Schrödinger equation. It is required that the coefficients $\Gamma_{k,l}$ constitute a matrix with only non-negative eigenvalues, and the $\hat{a}$-operators may be taken to be traceless.

Equation 7.37 is the famous *GKLS equation*, named after Vittorio Gorini, Andrzej Kossakowski, Göran Lindblad and George Sudarshan. Actually, we often see Eq. (7.37) referred to as just the *Lindblad equation* – probably because his proof is slightly more general than that of GKS, whose picture is shown in Fig. 7.2.

Figure 7.2 This photo shows GKS, or rather, with *permutation* corresponding to the photo, KSG, in 1975. Used with the permission of Springer Nature BV, from Ref. [14], permission conveyed through Copyright Clearance Center, Inc.

So, what did they prove? They proved that in order to ensure that the density matrix $\hat{\rho}$ maintains Hermicity, complete positivity[16] and trace, it *must* follow an evolution dictated by Eq. (7.37).

Now, why is it important to conserve trace and positivity?

7.2.4 Exercise: Preserving Trace and Positivity

In the following we will drop the subscript A on our reduced density matrix. We will also lose the hat, $\hat{\rho}_A \to \rho$.

In Section 2.6 we explained how the probability of measuring the outcome a for a measurement of some physical variable A is $|\langle\varphi_a|\Psi\rangle|^2$ where φ_a is the eigenstate of the operator $\hat{A}$ corresponding to eigenvalue a. For a density matrix ρ the corresponding probability is the diagonal element $\langle\varphi_k|\rho|\varphi_k\rangle$.

(a) Check that these two probability expressions are, in fact, identical in the case of a pure state, $\rho = |\Psi\rangle\langle\Psi|$.

(b) In order for a density matrix to produce sensible results, we must insist that it is positive semi-definite. A positive semi-definite matrix M has only non-negative eigenvalues, which, in turn is equivalent to requiring that

$$\mathbf{x}^\dagger M\mathbf{x} \geq 0 \tag{7.38}$$

for any vector $\mathbf{x}$. Correspondingly, a positive semi-definite operator $\hat{A}$ fulfils

$$\langle\psi|\hat{A}|\psi\rangle \geq 0 \tag{7.39}$$

for any state ψ.

[16] The notion of *complete* positivity is somewhat more strict than that of just positivity. However, will not enter into this distinction here.

Why must we insist that ρ has this property? And how can we even be sure that expressions such as the left hand side of Ineq. (7.39) are real in the first place?

(c) As discussed in Section 3.3, the set of all possible normalized eigenstates of a Hermitian operator, $\hat{A} = \hat{A}^\dagger$, forms an orthonormal basis – or, at least, can be *constructed* to form an orthonormal basis. We have also learned that the probabilities of all possible outcomes must sum up to 1.[17] Explain how this leads to the requirement that the trace of a density matrix must be 1 at all times.

(d) As any matrix, the density matrix can always be written in terms of a *singular value decomposition*. Since it is also Hermitian, this decomposition becomes particularly simple:

$$\rho = \sum_n p_n |\psi_n\rangle\langle\psi_n|, \tag{7.40}$$

where the $|\psi_n\rangle$ are orthonormal and the coefficients p_n are real.

The positivity and trace requirements on ρ impose restrictions on the p_n. Which ones?

Although not strictly necessary here, it may be useful to know that the identity operator may be expressed as

$$\hat{I} = \sum_n |\alpha_n\rangle\langle\alpha_n|, \tag{7.41}$$

where $\{|\alpha_n\rangle\}$ is any orthonormal basis for the space in question.

So, in order for a density matrix to produce physical predictions that make sense, it *must* be positive semi-definite and have trace 1 at all times.

The singular value decomposition of Eq. (7.40) illustrates how density matrices generalize the notion of wave functions. The special case of a wave function, or a *pure state*, emerges when all but one of the weights p_n are zero. The linear combination of Eq. (7.40) is *very* different from any linear combination of *wave functions*,

$$|\Psi\rangle = \sum_n a_n |\psi_n\rangle, \tag{7.42}$$

perhaps more different than a swift comparison between the above equation and Eq. (7.40) would suggest. In general, a linear combination of wave functions is put together by amplitudes a_n – complex amplitudes that carry both a magnitude and a phase. This, in turn, allows for interference, as we have seen in Exercises 2.3.3 and 5.3.3, for instance. The p_n's, on the other hand, are *classical probabilities*, they are all real and non-negative. No interference is brought about by the mixedness of Eq. (7.40). In this sense, Eq. (7.40) may be thought of as a classical ensemble of quantum wave functions.

The interaction between a quantum system and its surroundings will in general cause the quantum traits of the system to diminish. One manifestation of this is the issue of *decoherence*, which, as mentioned, is a huge challenge when it comes to building

[17] This is why the square modulus of a spatial wave function must integrate to 1 and why $|a|^2 + |b|^2$ in Eq. (4.7) must be 1.

working quantum computers. As we discussed in Section 6.5, the coherent nature of a quantum system, the one that allows for interference, is crucial for a quantum computer to provide any advantage over classical ones. However, in real life, the state of a quantum computer will eventually get entangled with its environment – with the consequence that its density matrix loses purity. It becomes less quantum-like and more classical-like.

We will use the GKLS equation, Eq. (7.37), to study one particular source of decoherence, one that is referred to as *amplitude damping*.

7.2.5 Exercise: A Decaying Quantum Bit

In Section 6.4 we introduced the notion of qubits, which are linear combinations of two states labelled $|0\rangle$ and $|1\rangle$, Eq. (6.18). In physical implementations of quantum computers, it is hard keep both states stationary; $|1\rangle$ will typically 'fall down' into $|0\rangle$ spontaneously – at a certain rate which we want to keep as low as possible.

Now, we could try to model this by introducing effective interactions into the Schrödinger equation. It wouldn't work, though:

(a) Suppose we start with the Hamiltonian of Eq. (5.3) with $W = 0$. Now, the upper right element in the Hamiltonian would induce transitions from $|1\rangle$ to $|0\rangle$. Why is it not admissible to insert a non-zero matrix element here while keeping the other off-diagonal element zero,

$$H = \frac{1}{2}\begin{pmatrix} -\epsilon & 0 \\ 0 & \epsilon \end{pmatrix} \to \frac{1}{2}\begin{pmatrix} -\epsilon & \Gamma \\ 0 & \epsilon \end{pmatrix}? \tag{7.43}$$

Suppose we did it anyway and solved the Schrödinger equation; what consequences would this have for the evolution of the wave function?[18]

So, spontaneous, irreversible transitions, such as the one we want to implement, cannot happen in any evolution governed by a Schrödinger equation with a Hermitian Hamiltonian. However, the GKLS equation, Eq. (7.37), allows us to introduce it. We impose a single $\hat{a}$-operator, a so-called *jump operator*, which turns $|1\rangle$ into $|0\rangle$:

$$\hat{a} = |0\rangle\langle 1| \quad \text{so that} \quad \hat{a}\,|1\rangle = |0\rangle. \tag{7.44}$$

With our usual representation in terms of vectors in $\mathbb{C}^2$, this projection operator may be written

$$\hat{a} = |0\rangle\langle 1| = \begin{pmatrix} 1 \\ 0 \end{pmatrix}\begin{pmatrix} 0 & 1 \end{pmatrix} = \begin{pmatrix} 0 & 1 \\ 0 & 0 \end{pmatrix}. \tag{7.45}$$

Thus, it is non-zero for the same matrix element as the one we discussed above, in regard to Eq. (7.43). But this jump operator will not enter into the Hamiltonian.

(b) With the Hamiltonian of Eq. (5.3), write out the GKLS equation, Eq. (7.37), with the single projection operator a above in the last term on the right hand side of

[18] You are welcome to check this out numerically.

Eq. (7.37) – combined with the single positive parameter Γ. In this case, there will be no summation over k and l since there is only one term. Write the equation for each of the four elements of ρ separately. This, in turn, constitutes a set of four coupled ODEs. In fact, it has the form of the Schrödinger equation – with a non-Hermitian effective Hamiltonian.

(c) In the case that $W = 0$, determine the diagonal elements ρ_{00} and ρ_{11} as functions of time with the initial condition

$$\rho(t=0) = |1\rangle\langle 1| = \begin{pmatrix} 0 & 0 \\ 0 & 1 \end{pmatrix}. \tag{7.46}$$

How does the purity, γ of Eq. (7.31), evolve as a function of time in this case?

(d) Now, with a non-zero W in the Hamiltonian of Eq. (5.3) and real, positive values for the parameters ϵ and Γ, solve the GKLS equation. As usual, we suggest that you do it numerically. Compare your solution with the unitary one, the one with $\Gamma = 0$, which you arrived at in Exercise 5.1.1, for various values of Γ. Use the same initial condition as in (c).

(e) As the master equation does not feature explicitly time-dependent terms, it will converge towards a *steady state*; eventually, in the limit $t \to \infty$, the density matrix ρ will become constant. Set the time derivative of ρ in Eq. (7.37) equal to zero and determine this state – numerically.

(f) In Exercise 6.4.5 we found that the NOT gate could be implemented by setting $\epsilon = 0$ and fixing the duration T at $T = \pi\hbar/W$. Suppose we expose such a system to the amplitude damping we have now implemented. If we now, starting out in the (pure) $|0\rangle$ state, $\rho(t=0) = |0\rangle\langle 0|$, implement the NOT gate from Exercise 6.4.5 with a positive decay rate Γ, what is the probability of actually reading off 1? In other words, play around with different values for Γ and determine $\rho_{1,1}(T)$. For simplicity, set $W = 1$.

The decay is not only manifested in the fact that the system is pulled towards the $|0\rangle$ state, it also introduces a damping of the oscillations we would have for an isolated quantum system undergoing unitary evolution. Note also that as long as $W \neq 0$ the system will, despite the pull towards $|0\rangle$, maintain a finite population of the $|1\rangle$ state even as t approaches infinity. These aspects are illustrated in Fig. 7.3.

In this model, we simply imposed an irreversible, spontaneous decay by hand – mathematically. Physically, it could be induced via the interaction with some background radiation field. We will follow the same *heuristic*[19] path in the last exercise of this chapter. It is a revisit of Exercise 2.4.1. In the last part of that exercise, we let our wave packet hit a well instead of a barrier by introducing a negative barrier height V_0 in the potential of Eq. (2.30). This is, more or less, the same system we studied in Exercise 3.1.1 – except that in that case the particle was trapped in the potential from the outset, in Exercise 2.4.1 it was unbound.

[19] A *heuristic* approach is one in which you allow yourself to simplify the problem and take mental shortcuts in order to arrive at a working solution in a more straightforward and less rigorous manner.

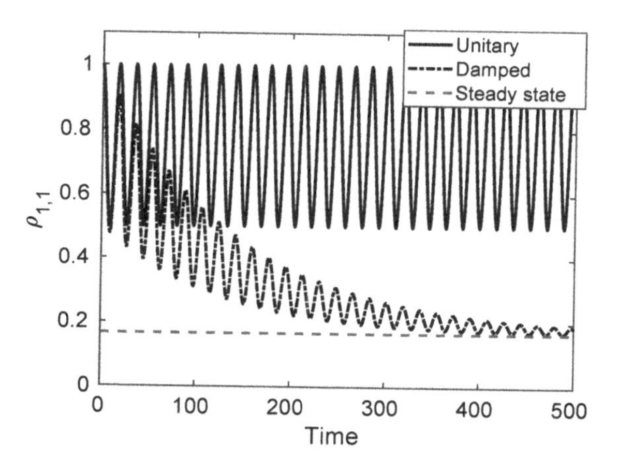

Figure 7.3 The full curve shows the probability of measuring a qubit to be in the $|1\rangle$ state as a function of time for a unitary evolution as dictated by the Hamiltonian of Eq. (5.47). The dash-dotted curve is the same, albeit with damping. This evolution is dictated by the GKLS equation, Eq. (7.37), with the jump operator of Eq. (7.44). The dashed curve is the probability of measuring the system in the $|0\rangle$ state for the steady state. This particular calculation corresponds to a calculation with the parameters $\epsilon = 0.5$, $W = 0.25$ and $\Gamma = 0.01$ in units where $\hbar = 1$. Clearly, this results in very strong decay.

For the free particle hitting the well, we found that a well would actually induce reflections – contrary to classical physics and, probably, also intuition. What we didn't discuss much was the possibility of *capture*. Isn't it possible for a particle to fall into the well and get trapped?

No, it is not – not in the context of the unitary evolution dictated by the Schrödinger equation. A particle with an energy that is positive from the outset could not get trapped without getting rid of energy somehow. However, with a time-independent Hamiltonian, there simply isn't any mechanism for taking away energy. This is in line with the beautiful theorem of Emmy Noether (pictured in Fig. 7.4), which states that for each symmetry in a physical system there is a conserved quantity. In this case, the conserved quantity in question is energy and the symmetry is time invariance. We touched upon another relevant example in the discussion following Exercise 2.6.2: when the system remains unaffected by a translation – when the Hamiltonian does not have any local potential – momentum is conserved.

There certainly exist ways of allowing for energy exchange for our quantum particle. One way could be to allow for time dependence in our Hamiltonian – as in Chapter 5. We could also let our particle interact with another system, such as a photon field. In this case, energy could be carried over from our quantum particle to the field. Or, possibly, in more familiar terms: the particle could get stuck in the well by emitting a photon which carries away the surplus energy. In that case, the total energy of the composite system would be conserved, but not the energy of the particle alone. In line with things we have discussed before, such a bipartite system is quite hard to describe

Figure 7.4 The German mathematician Emmy Noether (1885–1935) made several significant contributions to mathematics and mathematical physics. Most notable are her profound results on how symmetries lead to conserved quantities in physical systems.

as the photon field has very many degrees of freedom. We would need a very large computer and a whole lot of patience to describe the full system.

Instead, let's generalize our approach from Exercise 7.2.5.

7.2.6 Exercise: Capturing a Particle

When a quantum particle interacts with a photon field, or some other quantized field, the particle may decay from its initial state to another state with lower energy. It is often assumed that this happens spontaneously, with a certain probability rate r:

$$\frac{\mathrm{d}}{\mathrm{d}t} P_f(t) = r\, P_i(t), \qquad \frac{\mathrm{d}}{\mathrm{d}t} P_i(t) = -r\, P_i(t), \tag{7.47}$$

where $P_i(t)$ is the population of some initial state ψ_i, $P_i(t) = |\langle \psi_i | \Psi(t) \rangle|^2$, and P_f is the population of some final state ψ_f. Often, the *Fermi golden rule* can be used to predict this decay rate:

$$r_{i \to f} \sim \left| \langle \psi_i | \hat{H}_I | \psi_f \rangle \right|^2, \tag{7.48}$$

where $\hat{H}_I$, which is part of the Hamiltonian in question, is the interaction, or the *perturbation*, that induces the decay. Here, the total Hamiltonian, $\hat{H} = \hat{H}_0 + \hat{H}_I$, is time independent, and the interaction term $\hat{H}_I$ is assumed to be relatively weak. Usually, both ψ_i and ψ_f are taken to be stationary states of the unperturbed part of the Hamiltonian $\hat{H}_0$. Here, however, we allow ourselves to let the initial state Ψ_i evolve in time.

We will, for simplicity, assume that our system only has *one* bound state.

Next, label all the numerical eigenstates of our time-independent Hamiltonian $\hat{H}_0$ by φ_k – each with eigenenergy ε_k, where $\varepsilon_0 < 0$ and all other eigenenergies are positive. The numerical eigenstates with positive energies do not really represent physical states. First of all, these states should really constitute a continuum. Second, with our usual choice for approximating the kinetic energy, our FFT implementation, that is, they fulfil periodic boundary conditions, which makes them even less physical. Nonetheless, mathematically, we may still describe our physics in terms of these somewhat artificial states.

In the GKLS equation, Eq. (7.37), take the $\hat{a}_k$ operators, the jump operators, to bring about a jump from unbound state φ_k, $k > 0$, to the bound state φ_0:

$$\hat{a}_k = |\varphi_0\rangle\langle\varphi_k|. \tag{7.49}$$

Moreover, we define the Γ-coefficients in a manner inspired by Eq. (7.48):

$$\Gamma_{k,l} = \Gamma_0 \, \langle\varphi_0|x|\varphi_k\rangle \, \langle\varphi_l|x|\varphi_0\rangle, \tag{7.50}$$

where the interaction term $H_I \sim x$ has been inserted. This is in line with the interaction we saw in Eq. (5.27). Here, the proportionality factor Γ_0 determines the strength of the decay – and ensures appropriate units.

(a) For the same system as in Exercise 2.4.1, choose a set of parameters which is such that it has only *one* bound state. For this, the absolute value of your (negative) V_0 must be comparatively low.
(b) Analytically, set up the GKLS equation, Eq. (7.37), with the *dissipative* terms as dictated by Eqs. (7.49) and (7.50). Let the first term on the right hand side of Eq. (7.37), the commutator that provides the unitary, non-dissipative part of the evolution, be given by $\hat{H}_0$ alone.
(c) Assume that the solution of the time-dependent GKLS equation can be separated in two non-overlapping terms at all times:

$$\rho = \rho' + \rho_{0,0}|\varphi_0\rangle\langle\varphi_0|. \tag{7.51}$$

Show that the system decouples in separate equations for ρ' and $\rho_{0,0}$.

Also, explain why ρ' remains a pure state, $\rho' = |\Psi'\rangle\langle\Psi'|$, which follows a Schrödinger equation with an effective, non-Hermitian Hamiltonian.
(d) In the case that we start out in an unbound eigenstate, $\psi_i = \varphi_i$, is the rate at which $\psi_f = \varphi_0$ is populated consistent with Eq. (7.48)?
(e) Why can we be absolutely certain that the probabilities $\langle\Psi'(t)|\Psi'(t)\rangle$ and $\rho_{0,0}(t)$ add to 1 at all times?[20]

[20] You can assume that the $\Gamma_{k,l}$ of Eq. (7.50) constitutes a positive semi-definite matrix.

(f) At last, solve the GKLS equation for this particular system. This may be done conveniently by solving for Ψ' and the ground state probability $\rho_{0,0}$ separately. Use the same parameters as in part (a) for your potential, and let your initial state be a Gaussian that is well separated from this potential. With this in mind, set the parameters x_0, p_0, σ_p and τ as you please. Also, choose a comparatively low value for the decay strength Γ_0 in Eq. (7.50).

How does the capture probability $\rho_{0,0}(t)$ evolve in time? How is the final capture probability $\rho_{0,0}(t \to \infty)$ affected by your choices for p_0 – and τ?

In Fig. 7.5 we show a similar, albeit much more complicated, capture process. Here an initially free electron collides with an F^{6+} ion, a fluorine atom with six of its electrons stripped off. With its three remaining electrons, it resembles a lithium atom – with a strongly charged nucleus. The upper graph in the plot shows the probability for the incoming electron to be captured by the ion so that it forms a stable F^{5+} ion. As we can see, it has several pronounced peaks for specific energies of the incoming electron.

At the end of the day, we may be perceived as somewhat self-contradicting here; we argued that Eq. (7.43) had to be dismissed on account of not producing any unitary evolution, which, as we learned in Exercise 6.4.4, is related to the Hermicity of the Hamiltonian. And yet, in Exercise 7.2.6(c) we solved a Schrödinger equation with a non-Hermitian Hamiltonian. And it gets worse, the next chapter is dedicated to such Hamiltonians.

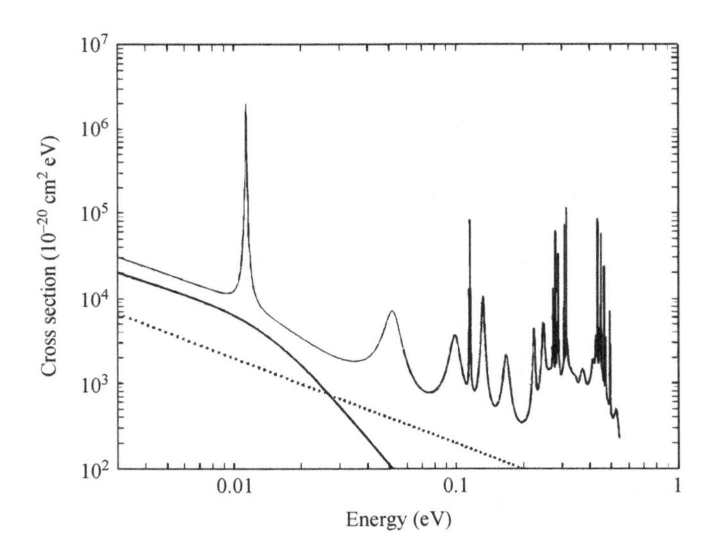

Figure 7.5 The upper graph shows the probability for an incoming, colliding electron to be captured by an F^{6+} ion as a function of energy. The sharp peaks correspond to a process called *dielectronic recombination*, which we will return to in the next chapter. Figure reprinted, with permission, from Ref. [38], M. Tokman, N. Eklöw, P. Glans, E. Lindroth, R. Schuch, G. Gwinner, D. Schwalm, A. Wolf, A. Hoffknecht, A. Müller, and S. Schippers, *Physical Review A* 66, 012703 (2002). Copyright (2002) by the American Physical Society.

A more precise reasoning for departing from the Schrödinger equation in Exercise 7.2.5 is the fact that it is unable to describe an irreversible process. The GKLS equation, on the other hand, is able to do so – through the last term on the right hand side, the *Lindbladian*. As we have seen, it can encompass irreversible processes such as spontaneous decay – accompanied with the loss, or *dissipation*, of energy. The fact that Noether's theorem does not apply here is also related to this; energy conservation applies to time-independent *Hamiltonians*, not to Lindbladians.

8 Non-Hermitian Quantum Physics

You may feel like objecting to the title of this chapter; aren't operators in quantum physics always Hermitian? Didn't we show in Exercise 1.6.3 that physical quantities would come out complex if not? The answer to both questions is *yes*. However, this chapter would be a very boring one if that were the whole story.

Actually, although not physical, allowing for non-Hermitian operators may be a very useful theoretical/numerical tool in many situations. Here we will take a brief look at a couple of such situations.

8.1 Absorbing Boundary Conditions

In Exercise 2.3.2 we saw that odd things happen to a wave packet approximation when it hits the boundary of a numerical domain. It could bounce back, as was the case in our finite difference implementations, or it could reappear on the other side of the grid, as we saw happening with our discrete Fourier transform approximation of the kinetic energy operator. Both of these are, of course, artefacts that come about because our numerical domain is too small. In order to avoid it, we must make sure we choose a domain large enough to actually contain the wave function. Or do we?

Let's consider the situation in Exercise 5.2.3, for instance, in which an initially bound system is partially liberated after being exposed to a laser field and outgoing, unbound waves are emitted. Several wave components, fast and slow ones, will escape the confining potential – at various times. Over the time it takes for the slower wave components to escape, fast components may have travelled quite far. In order to contain such an unbound wave function, we must apply a numerical domain that extends quite far, which, in turn, comes at a rather high price in terms of computation efforts in time and memory. This may be a bit frustrating if we really only want to describe what goes on in some smaller interaction region. Couldn't we rather kill off the outgoing parts when they approach the boundary, and focus on our attention on whatever is left?

Yes we could. There are several ways of doing this – of imposing *absorbing boundary conditions*. Arguably, the simplest one is to augment the Hamiltonian with an extra, artificial potential:

$$\hat{H} \to \hat{H}_{\text{eff}} = \hat{H} - \mathrm{i}\Gamma(x), \tag{8.1}$$

where $\hat{H}$ is the proper, Hermitian Hamiltonian and $\Gamma(x)$ is a function that is zero on most of the domain and positive close to the boundary. We coin this artificial amendment to our Hamiltonian a *complex absorbing potential*.

8.1.1 Exercise: Decreasing Norm

(a) Show that $\hat{H}_{\text{eff}}$ in Eq. (8.1) is non-Hermitian.

(b) Show that, for any wave function that has a certain overlap with the absorbing potential Γ, the norm of the wave function is decreasing in time.

Now, with this machinery in order, we may use it to facilitate solving some of the exercises we have already done. Let's start off with a revisit of Exercise 5.2.3.

8.1.2 Exercise: Photoionization with an Absorber

Start off with the implementation you used to solve Exercise 5.2.3; we will not change it much. But this time we will allow ourselves the luxury of choosing an L value that is likely to be too small – too small in the sense that a significant part of the wave function would hit the boundary unless we remove it before it gets there. To that end we impose this complex absorbing potential:

$$\Gamma(x) = \begin{cases} \eta(|x| - x_{\text{onset}})^2, & |x| > x_{\text{onset}}, \\ 0, & \text{otherwise}, \end{cases} \tag{8.2}$$

where the parameter η is the strength of the absorber which is non-zero only for $|x|$ larger than x_{onset}. In a practical implementation, you augment your Hamiltonian with this complex absorbing potential simply by adding $-i\Gamma(x)$ to the actual potential $V(x)$.

(a) Add this artificial contribution to your potential in your implementation from Exercise 5.2.3.

When you run it this time, you can set your box size $L = 100$ length units; this is a reduction by a factor four. To have an equally dense numerical grid as in Exercise 5.2.3, make the same reduction in the number of grid points.

For the complex absorbing potential, you can set $\eta = 0.01$ and $x_{\text{onset}} = 30$ length units.

When you run – and visualize – your simulation, pay particular attention to the evolution of your wave packet in the region where absorption goes on – for $x < -x_{\text{onset}}$ and $x > x_{\text{onset}}$.

(b) During time evolution, investigate how the norm N of the wave function,

$$N^2(t) = \langle \Psi(t) | \Psi(t) \rangle, \tag{8.3}$$

changes and plot N^2 as a function of time after having done your simulation.

(c) In order for your absorber to 'swallow' all the outgoing waves, you will need to propagate for a while longer than in Exercise 5.2.3; you probably need to increase your ΔT from back then. When you run your calculation long enough for the norm to reach a constant value, does the loss in norm, $1 - N^2(t \to \infty)$, coincide with the ionization probability you found in Exercise 5.2.3? Should it?

(d) Play around a bit with the absorber strength η in Eq. (8.2). There is a window of reasonable values. However, you do run into trouble if it is either too high or too low. What kind of trouble is this?

The answer to the question in (c) is *yes* – provided that we propagate long enough and the artificial complex absorbing potential works as it should, swallowing everything that hits it. This need not always be the case, though. If the absorbing potential is too strong, it will reflect parts of the outgoing waves – the slow ones in particular. For this reason, and the mere fact that low-energy wave components struggle to actually reach the absorber within reasonable time, the ionization probability predicted in this way may come out a little bit too low in practice.

If the absorber is too weak, on the other hand, parts of the wave will make it through the absorption region, hit the numerical boundary and, thus, be subject to the artefacts we wanted to avoid in the first place. A complex absorbing potential typically works better the larger the region on which it is allowed to act. However, a larger absorption region would require a larger numerical box, which, in turn, reduces the advantage of using a smaller numerical domain. So there is a trade-off here.

Although there is, to some extent, a need for optimizing the parameters of a complex absorbing potential, such techniques do remain useful tools when studying unbound quantum systems. It is quite common to include them when studying ionization phenomena, for instance.

While we're at it, why not apply our complex absorbing potential to the scattering example we studied in Exercise 2.4.1 as well?

8.1.3 Exercise: Scattering with an Absorber

(a) Rerun your simulation from Exercise 2.4.3 with the same initial set of parameters, which, apart from $V_0 = 1$, are listed in Exercise 2.4.1. However, this time, introduce the complex absorbing potential in Eq. (8.2). Again, seize the opportunity to reduce your box size L considerably. Let's halve it – and the number of grid points $n + 1$ as well. For the absorber, you may set $\eta = 0.01$, as in Exercise 8.1.2, and $x_{\text{onset}} = 40$.

Check, by direct inspection as you run your simulation, that your parameters η and x_{onset} are such that virtually no reflection goes on near the boundaries. Adjust the parameters if necessary.

Instead of fixing the duration of your simulation, it may be more convenient to set it running until it is more or less depleted – until the square of the norm, Eq. (8.3), is, say, less than 1%.

(b) Now, instead of determining reflection and transmission probabilities as in Eqs. (2.32), let's just measure the amount of absorption at each end:[1]

$$R \approx \frac{2}{\hbar}\int_0^\infty \int_{-L/2}^{-x_{\text{onset}}} \Gamma(x)|\Psi(x;t)|^2 \,\mathrm{d}x\,\mathrm{d}t = \frac{2}{\hbar}\int_0^\infty \langle \Psi(t)|\Gamma_{\mathrm{L}}|\Psi(t)\rangle \,\mathrm{d}t, \tag{8.4a}$$

$$T \approx \frac{2}{\hbar}\int_0^\infty \int_{x_{\text{onset}}}^{L/2} \Gamma(x)|\Psi(x;t)|^2 \,\mathrm{d}x\,\mathrm{d}t = \frac{2}{\hbar}\int_0^\infty \langle \Psi(t)|\Gamma_{\mathrm{R}}|\Psi(t)\rangle \,\mathrm{d}t, \tag{8.4b}$$

where we have split the absorbing potential, Eq. (8.2), into a left and a right part:

$$\Gamma_{\mathrm{L}} = \gamma(-x_{\text{onset}} - x) \text{ for } x < -x_{\text{onset}} \quad \text{and} \quad \Gamma_{\mathrm{R}} = \gamma(x - x_{\text{onset}}) \text{ for } x > x_{\text{onset}}. \tag{8.5}$$

Do you arrive at the same transmission and reflection probabilities as last time? Feel free to check for various values of the momemtum p_0 of the incoming wave.

As mentioned, it is quite common to impose absorbing boundary conditions like this – or in some similar way – when simulating the time evolution of unbound dynamical systems. As we will see in the following, techniques involving non-Hermitian Hamiltonians are frequently used in time-independent contexts as well. We will, however, not leave our scattering implementation just yet.

8.2 Resonances

In Section 3.1 we claimed that the spectrum, the set of all eigenvalues, of a Hamiltonian will generally consist of a discrete set corresponding to bound states and a continuous set, which corresponds to unbound states. That would have been the whole story if it hadn't been for the *resonance states*.

8.2.1 Exercise: Scattering off a Double Well

We continue to revisit old exercises, this time Exercises 2.4.4 and 6.1.1. We expose our quantum particle to the same double barrier as in the former, Eq. (2.36), but now we let our particle scatter off it, as in Exercises 2.4.1 and 8.1.3, instead of placing it between the barriers initially, as we did in Exercise 2.4.4.

(a) Simulate a scattering event in which you replace the single barrier, Eq. (2.30), with the double barrier of Eq. (2.36). The parameters to use for the double barrier are listed here, along with a number of other physical parameters and a few suggestions for the numerical ones:

[1] The factor 2 in front of the integrals may come across as somewhat non-intuitive. However, you may arrive at these formulae by first writing out the evolution dictated by the non-Hermitian version of the von Neumann equation, Eq. (7.25): $i\hbar\,\partial/\partial t\,\rho = H\rho - \rho H^\dagger$, and then accumulate in time the trace of the absorbed part in the left and right ends, respectively.

L	$n+1$	Δt	η	x_{onset}	σ_p	p_0	x_0	τ	V_0	w	s	d
100	1024	0.1	0.05	40	0.1	1	−20	0	4	3	25	3

You may note that this amounts to a much denser numerical grid than what we used in the previous exercise. This is necessary because of the narrow barriers and their rather sharp corners.

As in Exercise 8.1.3, run your simulation and determine the transmission probability.

(b) Now, repeat this calculation for a range of initial mean momenta or, rather, mean energies $\varepsilon_0 = p_0^2/(2m)$, ranging from 1 energy unit up to $V_0 = 4$ units. Do not bother to simulate your wave packet on the fly. Just use Eq. (8.4b) to determine the transmission probability T as a function of energy ε_0 and plot it.

It's not very monotonic, is it? Can you pinpoint specific energies ε_0 for which the transmission probability behaves peculiarly?

(c) Hopefully, you found that $T(\varepsilon_0)$ has pronounced peaks at certain ε_0 values. Actually, these peaks would be even more pronounced if it hadn't been for the fact that our initial wave has a finite width σ_p in momentum. With a sharper momentum distribution, we would get sharper peaks.

The underlying structure we are trying to get at would be revealed if we could extrapolate the momentum width σ_p to zero somehow. This may appear difficult as it would require that the spatial width of our wave would approach infinity – as dictated by Ineq. (1.5). However, we are quite capable of working around this. We already did so in Exercise 6.1.1. There the incoming wave was a pure exponential of the form $\exp(ip_0x/\hbar)$, which, in turn, corresponds to $\sigma_p = 0$.

In your implementation from Exercise 6.1.1, shift your double barrier a bit towards the right so that you may quite reasonably assume that it is zero at $x = 0$ and below and at $x = D$ and beyond – for some D-value larger than $2d + w$. Next, adjust and apply your implementation from Exercise 6.1.1 to determine $T(\varepsilon_0)$ in the limit $\sigma_p \to 0$. Do this for ε_0 ranging from almost zero to V_0.

How does this $T(\varepsilon_0)$ compare to the preceding one?

(d) What would the corresponding $T(\varepsilon_0)$ look like if there were only one barrier? Just rerun your calculation from (c) with a single barrier.

So what are these sharp peaks that emerged in our plot for transmission through the double well – the ones you may see in the upper panel of Fig. 8.2? They are manifestations of *resonances*. This particular type are called *shape resonances*. We will try to provide ways of understanding how they come about.

But before we do that, let's take a moment to dwell on what we just saw. It looks rather odd seen through classical glasses. Suppose you follow our incident wave and hit a single barrier. As we have seen, for comparatively low energies, most of the wave will be reflected. However, if a second barrier is present behind it, there may be a much higher chance your wave will just go straight through; at the right energy, virtually all of the wave will pass. But how could the initial wave even 'know' that there was a second barrier in the first place, as most of the wave would have bounced at the first barrier and never reached the second? It is fair to call this a manifestation of the non-local, and non-intuitive, nature of quantum waves.

You may wonder what this has to do with resonance phenomena such as the one we studied in Exercise 6.3.1 – after all, the phenomena share the same name. Perhaps somewhat misleadingly, the name is motivated by the strong similarity between peaks such as the ones seen in the upper panel of Fig. 8.2 and in Fig. 6.6.

You may also wonder what all this has to do with non-Hermitian quantum physics. In this system, non-Hermicity emerges when we impose *outgoing boundary conditions*.

8.2.2 Exercise: Outgoing Boundary Conditions

This time we will lean heavily on Exercise 3.1.1. We will continue with our double well and require that the energy $\varepsilon < V_0$. But now the barriers will not be smooth but rather sharply rectangular. Apart from this, which corresponds to letting s approach infinity in Eq. (2.30), let the parameters of the potential be the same as in Exercise 8.2.1.

As illustrated in Fig. 8.1, the x-axis may be divided into five regions, I–V. The solution of the time-independent Schrödinger equation is a linear combination of $\exp(\pm ikx)$ in regions I, III and V and $\exp(\pm \kappa x)$ in regions II and IV where

$$k = \frac{1}{\hbar}\sqrt{2m\varepsilon}, \tag{8.6a}$$

$$\kappa = \frac{1}{\hbar}\sqrt{2m(V_0 - \varepsilon)}. \tag{8.6b}$$

(a) All in all, this should leave you with no less than 10 coefficients to determine in order to solve the time-independent Schrödinger equation. However, this time we only allow outgoing waves in regions I and V – waves that move away from the double barrier, that is. If you combine the space-dependent parts with the time factor $\exp(-i\varepsilon t/\hbar)$, you should be able to identify which coefficients should be set to zero.[2]

(b) As before, we simplify our problem by dealing with the symmetric and the anti-symmetric cases separately. If we start out with the former, we may set up our wave function as

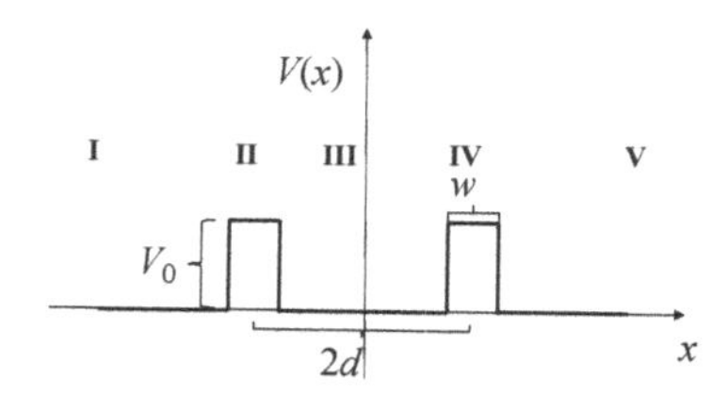

Figure 8.1 The potential under study in Exercise 8.2.2. The domain is divided into five sections – each with its own analytical expression for the solution of the time-independent Schrödinger equation.

[2] In case you feel like objecting to the fact that we insist on outgoing waves with nothing coming in, you are absolutely right to do so. Let's do it anyway.

$$\psi(x) = \begin{cases} A\cos(kx) & \text{in region III,} \\ Be^{-\kappa x} + Ce^{\kappa x} & \text{in region IV,} \\ De^{ikx} & \text{in region V,} \end{cases} \tag{8.7}$$

where the solutions in regions I and II are given by the fact that $\psi(-x) = \psi(x)$.

Explain why our four coefficients need to fulfil[3]

$$M \begin{pmatrix} A \\ B \\ C \\ D \end{pmatrix} = \begin{pmatrix} 0 \\ 0 \\ 0 \\ 0 \end{pmatrix} \quad \text{where} \tag{8.8}$$

$$M = \begin{pmatrix} \cos(kd_-) & -e^{-\kappa d_-} & e^{\kappa d_-} & 0 \\ -k\sin(kd_-) & \kappa e^{-\kappa d_-} & -\kappa e^{\kappa d_-} & 0 \\ 0 & e^{-\kappa d_+} & e^{\kappa d_+} & e^{ikd_+} \\ 0 & -\kappa e^{-\kappa d_+} & \kappa e^{\kappa d_+} & ike^{ikd_+} \end{pmatrix},$$

and we have introduced $d_\pm \equiv d \pm w/2$ for convenience.

In Exercise 3.1.1 we insisted that the determinant of a similar, albeit smaller, matrix had to be zero. We must do so again here. Why is that, again?

(c) This time you will not succeed in finding any real ε that gives $\det M = 0$. But you may be able to find some *complex* ones. For reasonably dense grids for both $\mathrm{Re}\,\varepsilon$ and $\mathrm{Im}\,\varepsilon$, run over several complex ε values in search of a value that makes the determinant of M vanish. Let the real part range from zero to V_0 and the imaginary part from -0.2 energy units to zero. When you plot the absolute value of $\det M$ as a function of $\mathrm{Re}\,\varepsilon$ and $\mathrm{Im}\,\varepsilon$, perhaps you arrive at something like the middle panel of Fig. 8.2?

(d) Repeat the same calculation for the case of an anti-symmetric wave function $\psi(-x) = -\psi(x)$. Do your findings resemble the lower panel of Fig. 8.2?

What happened there? Did we not learn in Exercise 1.6.3 that Hermitian operators, such as our Hamiltonian, will provide real expectation values and eigenvalues? The answer is that the momentum operator, which appears in the Hamiltonian, isn't actually Hermitian here. We gave that up when we allowed for 'wave functions' that blow up exponentially as $|x| \to \infty$. Our exponential solution in region V with a complex wave number $k = k_R - ik_I$, with positive k_I, is proportional to

$$e^{ikx} = e^{ik_R x} \cdot e^{+k_I x}. \tag{8.9}$$

This function will grow beyond all measure when x becomes large. The same happens at the other end, in region I. So, in fact, there is no contradiction with what we learned in Exercise 1.6.3(a). In order to prove that $\hat{p}^\dagger = \hat{p}$, we made direct use of the fact that

[3] Of course, the four equations could be formulated and ordered differently, resulting in a different but equally adequate matrix M.

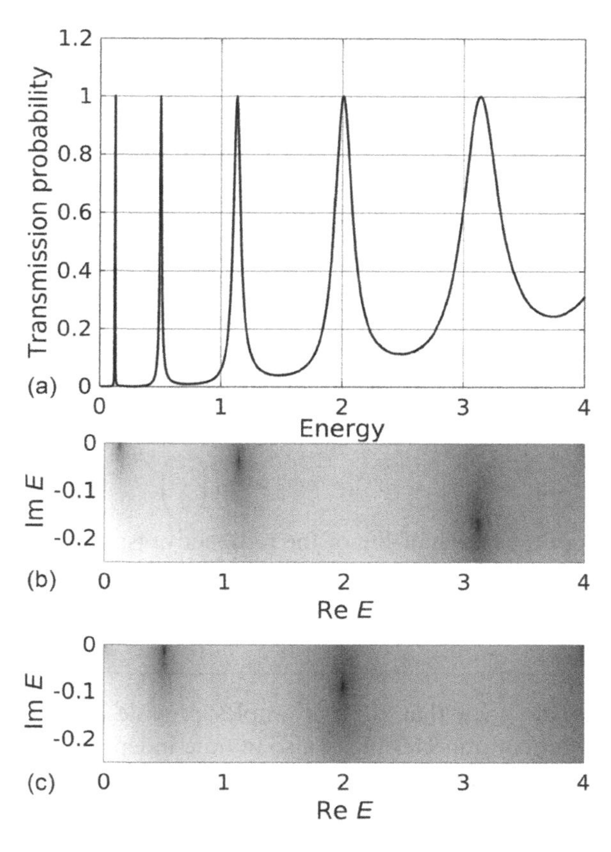

Figure 8.2 (a) The transmission probability as a function of the energy of an incoming wave for the system under study in Exercise 8.2.1. (b) The logarithm of the absolute value of $\det M$ in Eq. (8.8) for complex energies. The dark spots, where $\det M = 0$, represent symmetric solutions of the time-independent Schrödinger equation with outgoing boundary conditions. (c) The same as (b), however with anti-symmetric 'solutions'.

proper wave functions are to vanish for large values of x. If we extend our function space and allow $\hat{p} = -i\hbar\,d/dx$ to act on functions without this trait, $\hat{p}$ is no longer Hermitian.

Although these are not proper eigenenergies of any proper Hamiltonian, we do hope that the correspondence with the peaks we saw in Exercise 8.2.1 and the complex eigenvalues we just saw has not gone unnoticed. In Fig. 8.2 we try to make the connection so clear that it simply cannot be ignored. Note that the coincidence is not limited to the real part of the resonance 'energies' and the position of the peaks in $T(\varepsilon_0)$. Hopefully, it is almost equally clear that the imaginary part has something to do with the width of the peaks. And herein lies the advantage of allowing for complex energies: it provides us with both the position and the width of *resonance peaks*.

This width, in turn, is related to the so-called *lifetime* of the resonance.

8.2.3 Exercise: The Lifetime of a Resonance

Suppose that a quantum system starts out in a resonance state $\psi_{\rm res}$ with a complex eigenenergy given by

$$\varepsilon_{\rm res} = \varepsilon_{\rm pos} - {\rm i}\,\frac{\Gamma_{\rm width}}{2}, \tag{8.10}$$

where $\varepsilon_{\rm pos}$ and $\Gamma_{\rm width}$ are real numbers, and $\Gamma_{\rm width}$ is positive. The notion of *being in such a resonance state* may not be very well defined from a physical point of view, but for now, let's just assume that it makes sense.

Under this assumption, show that the square of the norm of the wave function, $\langle \psi_{\rm res}(t)|\psi_{\rm res}(t)\rangle$, will decrease exponentially.

Note that this is, in fact, the decay law for a radioactive material:

$$n(t) = n_0 \left(\frac{1}{2}\right)^{t/t_{1/2}} = n_0\, e^{-rt}, \tag{8.11}$$

where $t_{1/2}$ is the half-life of the radioactive isotope, r is the decay rate, see Eq. (7.47), and n_0 is the initial amount of radioactive material. We can set n_0 to 1 here.

How do the half-life $t_{1/2}$ and the rate r relate to the imaginary part of the energy, $\Gamma_{\rm width}$?

We do hope that these examples provide convincing arguments for sometimes allowing for non-Hermicity also in time-independent quantum physics – for practical reasons. However, what *are* resonance states? We have learned that eigenstates of a Hamiltonian could either belong to a discrete set, corresponding to localized, bound states, or to a continuous set, corresponding to unbound, unlocalized eigenstates. Do resonances actually represent a third way between the two? Not really, to the extent that resonance states actually are *states*, they belong to the continuum.

From a Hermitian point of view, resonances manifest themselves by the fact that several eigenenergies in the continuum pile up in a narrow energy region. From such pile-ups you may construct linear combinations which resemble bound states in the sense that they are more or less localized in space – and in energy. However, such a linear combination would not actually be a stationary solution; it is not a solution of the time-independent Schrödinger equation since the energy is not fixed. Although the energy standard deviation, or *width*, can be quite low, it is not zero, see Exercise 2.6.1. Due to this, a system starting out in such a state will not remain in it. And the probability of measuring it in the same state at a later time decreases in time. The wider the distribution in energy, the faster the decay. This decrease will often, to a good approximation, follow an exponential function in time with the rate given by the imaginary part of the resonance energy – as in Exercise 8.2.3.

An example of this is illustrated in Fig. 8.3. It corresponds to an initial Gaussian wave packet placed between two barriers as illustrated in Fig. 8.1. As we learned in Exercise 2.4.4, more and more of it will tunnel through the barriers and escape as time goes by. The figure shows the decreasing probability of remaining between the barriers. After a short while, the wave function is, between the barriers, very similar to the

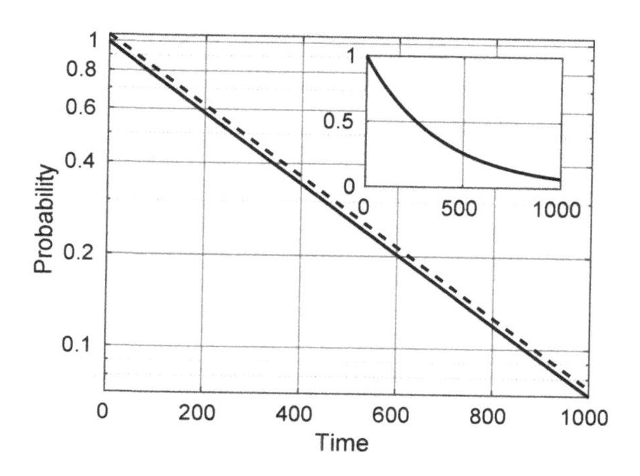

Figure 8.3 Here we have placed an initially localized wave packet between the two barriers identical to the ones under study in Exercise 8.2.2. The full curve shows the probability for the system to remain between the barriers as a function of time. The dashed curve is the exponential decay as dictated by the imaginary part of the complex energy of the first resonance seen in Fig. 8.2, the one with the narrowest, leftmost peak. To be able to distinguish it from the full curve, we have lifted it a bit. The insert shows the probability as a function of time with a linear y-axis.

resonance with the longest lifetime – or, correspondingly, the smallest width. The fact that the probability follows a straight line with a logarithmic y-axis goes to show that it is, indeed, exponential – consistent with what we discussed in Exercise 8.2.3.

While allowing for non-Hermicity is quite convenient when studying resonance phenomena in quantum physics, explicitly insisting on outgoing boundary conditions as in Exercise 8.2.2 is not the usual way of doing so in practice. Introducing explicitly anti-Hermitian terms in the Hamiltonian is more common. If this is done adequately, it enables us to calculate resonance states that abide by usual boundary conditions instead of blowing up, as in Eq. (8.9). Imposing a complex absorbing potential, Eq. (8.1), could, to some extent, be one way of doing so. However, methods that involve turning the position variable x into a complex-valued quantity has proven more practical [30]. The original and most straightforward way of implementing this is to simply multiply the position variable by a complex phase factor [7, 35].

Although the notion of resonances as eigenstates with complex eigenenergies for a non-Hermitian Hamiltonian is not really physical, the consequences of having resonances in the continuum is very real indeed.[4] One manifestation of this is shown in Fig. 7.5. As in the upper panel of Fig. 8.2, the peaks are due to resonance states. However, these resonances are different from the ones we saw in Exercise 8.2.2, these are *doubly excited states*.

[4] Pun intended.

8.2.4 Exercise: Doubly Excited States

We revisit a two-particle system similar to the one in Exercise 4.4.3. Here also the particles' interaction is given by Eq. (4.21), but this time we expose our particles to a Gaussian confining potential:

$$V(x) = -e^{-x^2/4}. \tag{8.12}$$

Moreover, we start off by ignoring the interaction between the two particles.

(a) Show that if we remove the interaction, if we set $W_0 = 0$ in Eq. (4.21), then product states of eigenstates of the one-particle Hamiltonian with the potential above, Eq. (8.12), would solve the two-particle time-independent Schrödinger equation.

Also, show that, if we impose the proper exchange symmetry or anti-symmetry on such product states, they are still solutions.

(b) With $W_0 = 0$ still, determine the energy of the spatially exchange-symmetric state in which both particles are in the first, and only, excited state.

This energy is actually embedded in the continuous part of the spectrum, which corresponds to an unbound system. Why?

Hint: What is the minimal energy for a system in which one particle is in the ground state and the other one is free and far away?

Suppose now that we gradually increase W_0 from zero. This will shift the energy of the doubly excited state of the, initially, non-interacting particles upwards. And, more importantly, it will mix this state with the 'true' continuum states. In a non-Hermitian context, such doubly excited states are identifiable with eigenstates of complex energy – with an increasing imaginary component of their 'energy'. This is illustrated in Fig. 8.4. This figure also illustrates the three bound states the system features when the interaction strength W_0 is comparatively low. The ground state is a spatially exchange-symmetric state in which both particles essentially are in the one-particle ground state. In the limit $W_0 \to 0^+$, both excited states correspond to one particle being in the excited one-particle state and the other in the one-particle ground state. This scenario comes in both an exchange-symmetric and an anti-symmetric flavour. For finite interaction strength W_0, the symmetric state is higher in energy.

The continuum of our two-particle system starts at the energy corresponding to the one-particle ground state – with an energy of $\varepsilon_0^{(1)} = -0.694$ energy units. This is the energy of a system in which one particle is in the ground state and the other is free – however, with very little energy. Our doubly excited state, the subject of Exercise 8.2.4, has an energy higher than this and, thus, it is embedded in the continuum. With the interaction turned on, the particles may exchange energy so that one may escape while the other relaxes to the one-particle ground state. This is an example of what is called *autoionization* – or the *Auger–Meitner effect*. Actually, this phenomenon is often referred as just the Auger effect – despite the fact that Lise Meitner, whose picture you can see in Fig. 8.5, discovered it first.

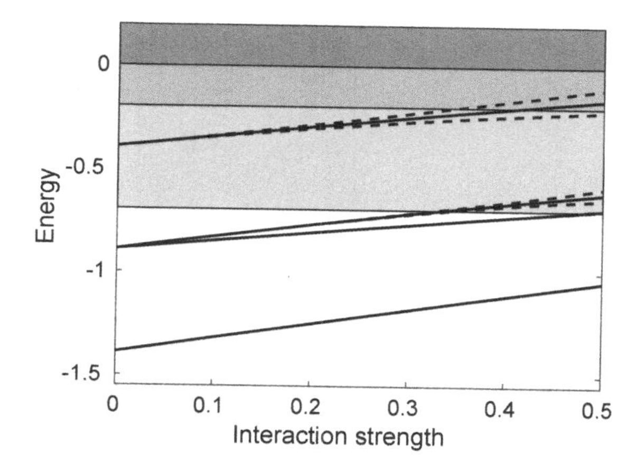

Figure 8.4 Here we have plotted energies for a two-particle system in which we have, quite artificially, adjusted the interaction strength, W_0 in Eq. (4.21), from zero up to 0.5 units. The four curves correspond to the possible combinations of the two bound one-particle states, which are eigenstates when $W_0 = 0$. The doubly excited state, which is embedded in the grey continuum, is a resonance state. The width, which is exaggerated for illustrative purposes, is indicated by dashed curves. With increasing W_0, the bound states also eventually reach the continuum and, in effect, become resonances. The darker grey area indicates the double continuum, where it is energetically possible for both particles to escape the confining potential, while the intermediate tone indicates the onset of the second ionization threshold , in which the remaining system is left in the excited state after emission of one particle.

Figure 8.5 Lise Meitner, who was born Austrian and later also acquired Swedish citizenship, and her collaborator Otto Hahn in their lab in 1912. Together they did pioneering work on nuclear fission. This work was awarded the 1944 Nobel Prize in Chemistry. However, the prize was given to Otto Hahn alone; Meitner was not included. She should have been.

The process under study in Fig. 7.5 is the time reverse of autoionization. The sharp peaks appear when an incoming, initially unbound electron enters with an energy such that the total energy of the system coincides with that of a doubly excited state. The capture probability is significantly increased when it is energetically possible for the

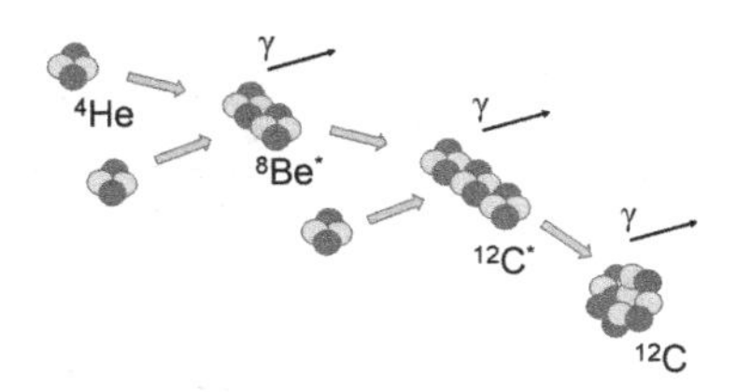

Figure 8.6 The ^{12}C nucleus, which consists of six protons and six neutrons, has a resonance state that corresponds to three ^{4}He nuclei, α*-particles*, coming together. The importance of this resonance state, which is called the *Hoyle state* after the British astronomer Sir Fred Hoyle, can hardly be overestimated as it enables the formation of stable carbon in stars [21]. In the process illustrated above, two α-particles collide to form an unstable ^{8}Be nucleus. This nucleus, in turn, fuses with yet another α-particle, forming a ^{12}C nucleus in the Hoyle state. This resonance may, in turn, relax to the stable ^{12}C ground state. In each step of this process, excess energy is emitted in the form of γ-radiation, a high-energy photon.

total system to form such a state, before it subsequently relaxes into an actual bound state by emission of a photon.[5]

In Fig. 8.6 a very similar – and important – process observed in nuclear physics is illustrated.

[5] The thicker curve in Fig. 7.5 is the contribution from a single, particularly wide resonance, while the dotted curve corresponds to the capture probability we would have without resonances.

9 Some Perspective

After having worked your way through this book, please do not go around telling your friends that you understand quantum physics. No one does.[1] But you have gained some acquaintance with how quantum physics works – by *doing* some quantum physics. You should still be careful bragging about it – not only because chances are that rather few people will actually be impressed, but also because virtually all of the examples and exercises you have seen in this book are quite simplified. That is not to say that they are all *easy*! Some of them are quite involved, and you should give yourself credit for managing to work your way through them. But we should also be aware that most of them pertain to single-particle systems; we know by now that many-particle systems are way more complicated. Moreover, most of the spatial wave functions we have looked at, both dynamic and stationary ones, reside in a one-dimensional world.

Another, related, issue is the fact that many exercises have been tailored in a manner that prevents certain difficulties from surfacing. It is a bit like playing curling; in preparing the exercises, we have quite often been running around with our brooms to clear the ice of obstacles that otherwise would have distorted and complicated the path of our stone. When solving a real world problem from a specific discipline pertaining to quantum physics from scratch, issues and problems we didn't foresee tend to emerge. It is rarely as straightforward as some of the exercises you have seen potentially would have you believe – although some of them certainly have been complicated enough.

So, in case you decide to continue acquiring more profound knowledge of quantum physics, within fields such as solid state physics, quantum chemistry, atomic, molecular and optical physics, nuclear physics or particle physics, it is important to know that several unforeseen complications will have to be dealt with. So be prepared for such challenges and don't allow them to break your spirit when faced with them. And know that you do already have a toolbox full of useful computational and analytical tools – tools that may be expanded upon and prove quite applicable should you seek a deeper familiarity with the quantum world.

Whether this book has worked as a primer for further quantum indulgence or simply a way of acquiring a certain acquaintance with the quantum world, we still hope that, through all the technical details, equations and lines of code, an appreciation for the beauty of the theory has emerged. Beneath all the technical mess, there is order.

[1] At least according to Richard Feynman.

9.1 But What Does It Mean?

Throughout the book questions regarding *quantum foundations* and interpretations, questions such as 'what does this actually mean?', have lingered underneath the surface. And most times we have rather shamelessly evaded them.

Perhaps the majority of physicists would, if obliged to choose a view on how to interpret quantum physics, point towards the *Copenhagen interpretation*, which is attributable to Niels Bohr and his circle – a circle that included Werner Heisenberg and Wolfgang Pauli. While it seems hard to find a clear-cut definition of what the Copenhagen interpretation actually *is*, it does relate to the notion that the outcome of a measurement is fundamentally random. It does not make sense to talk about the position, momentum or energy of a quantum particle without actually measuring it. And upon measurement, a process that is irreversible, at least *in effect*, takes place – the collapse of the wave function.

In a 1989 column in *Physics Today* [29], David Mermin 'defined' the interpretation in the following way: 'If I were forced to sum up in one sentence what the Copenhagen interpretation means to me, it would be "Shut up and calculate!".' Admittedly, this book has been rather loyal to this 'interpretation'.

The question of what actually constitutes a measurement is certainly a justifiable one. In trying to answer it, we cannot avoid the inclusion of an interaction between our quantum system and some measuring device. And since a full description of the combined system of apparatus and quantum system is usually beyond a practical description, a theory of open quantum systems – density matrices and such – lends itself as the adequate framework for describing the process of measurement. Perhaps the Copenhagen interpretation is not the full story, just the starting point – that a deeper, more detailed understanding of the measurement process will add to or adjust our standard quantum interpretation.

But there are alternatives. If you look, you will find some, not many, interpretations of quantum physics that are fundamentally different. Arguably, the most interesting one is the *pilot wave* interpretation, attributable to Louise de Broglie and David Bohm. Mathematically, it gives the same predictions as standard quantum physics. But it allows for an interpretation in which quantum objects do have well-defined positions and momenta – also prior to measurement (see Fig. 9.1). This, however, comes at the price of other conceptual difficulties.

9.2 Quantum Strangeness

We continue to quote David Mermin [29], who, fortunately, refuses to settle for just calculating: 'But I won't shut up. I would rather celebrate the strangeness of quantum theory than deny it, because I believe it still has interesting things to teach us about how certain powerful but flawed verbal and mental tools we once took for granted continue to infect our thinking in subtly hidden ways'.

Figure 9.1 It's hard to tell the quantum score between Bohm United and Copenhagen FC. However, the Danish team does seem to be a bit ahead.

He goes on to point out two pitfalls. The first one has to do with the fact that the strange, uncertain and stochastic nature of the quantum world can, and has, deceived people into believing that it is all a haze, we can't really know anything for certain. However, quantum theory has enabled us to extract substantial knowledge to a measure we could hardly dream of – and to make quantitative predictions of unprecedented precision.

In combating this sentiment, on the other hand, there is a risk of attempting to trivialize quantum strangeness. Perhaps it could be paraphrased along these lines: 'Yes, it may come across as a bit of a mystery to us, just like electromagnetism did back in the day. However, this mysticism will certainly diminish with increasing familiarity'. Some would say that there is a need to de-mystify quantum phenomena, especially now with the so-called *second quantum revolution* going on. Perhaps. But, if we take this too far, we miss out. By disregarding the genuine quantum strangeness, or, again quoting Mermin [29], by 'sanitizing the quantum theory to the point where nothing remarkable sticks out', chances are that novel insights, both scientific and philosophical ones – in addition to technological solutions – pass us by.

And, beyond the realms of the mere physical world, we may miss out on an important realization or experience. What we do not know, what we have not seen before, may be different in ways we could not have conceived in the first place. Learning happens and understanding grows by exposing orselves to the unknown rather than projecting our own perceptions and categories onto it.

9.3 What We Didn't Talk About

While we have tried to span rather widely in order to try to give a taste of several different quantum flavours, the true span is, of course, much wider. And, of course, it goes a whole lot deeper than the present account, which could, somewhat exaggeratedly, be compared to rushing through the Louvre exhibitions on a motorbike.

One topic that has not been properly introduced is the notion of *field theory*. In relation to Eq. (2.5), it was mentioned that the electromagnetic field should not really be a simple function but rather an *operator* – one that involves creation and annihilation of *photons*, which are the mediators of electromagnetic interactions. We say that these particles are the quanta of the photon field; the electromagnetic field is *quantized*. For the nuclear forces seen in nature, other mediators exist – mediators or *particles* that constitute the quanta of their respective quantized fields. And the notion of quantized fields extends further. On a fundamental level, it also applies to the particles that make up matter, such as electrons. As briefly mentioned in Section 7.1, particles should really be described in terms of quantized fields rather than wave functions. This is not to say that the wave function formulations we have been dealing with are *wrong*, they may be derived from the underlying formalism as a more than adequate approximation in most cases. However, at high energies, when relativistic effects come into play in full, the field theoretical formulation is the proper framework.

Another topic we haven't mentioned is that of *perturbation theory*. It is also an important topic within quantum physics, albeit for more technical reasons. A full description of a quantum system, be it stationary or dynamic, is often hard to achieve. In many situations it is a viable path to start off with a similar system that we know how to handle and then treat the difference between our simplified system and the actual one as a small disturbance – a *perturbation*. For instance, in the case of an atom exposed to an electric field, Eq. (5.26), it may be useful to treat the interaction term, Eq. (5.27), as a small perturbation of the atom – a perturbation that can be accounted for by applying an iterative scheme. The first-order correction can be determined in a rather simple way by assuming that the population of the initial state is virtually unchanged; the interaction only leads to a minor change in the wave function. A second-order correction can, in turn, be calculated by taking the first-order correction as the starting point – and so on. If the electric field is comparatively weak, it may be sufficient to include only the first few corrections to get sensible results. In this way, quantities such as ionization and excitation probabilities may be approximated for rather complex systems without actually solving the time-dependent Schrödinger equation – even analytically in some cases.

When addressing the time-independent energy spectra of many-particle atoms and molecules, we have several times alluded to the somewhat artificial notion of non-interacting electrons. Such a system would correspond to Eq. (2.5) with both the electromagnetic field, given by the vector potential $\mathbf{A}$, and the electron–electron interaction W set to zero. This notion makes more sense in the context of perturbation theory, in which the remaining sum of single-particle Hamiltonians is considered the unperturbed system and the electron–electron repulsion, Eq. (2.7), may be treated as a perturbation.

In this context, the framework of *density functional theory* also deserves mention. It is a very much applied methodology – particularly within quantum chemistry and solid state physics. It aims to lift the curse of dimensionality by describing an N-particle quantum system by means of the density function,

$$n(x) = N \int_{-\infty}^{\infty} \cdots \int_{-\infty}^{\infty} |\Psi(x, x_2, \ldots, x_N)|^2 \, dx_2 \cdots dx_N, \tag{9.1}$$

instead of the full wave function. This is quite an ambition: for N particles in a d-dimensional world, it consists in reducing the problem from describing a complex function in an Nd-dimensional space to a real function in just d dimensions. While ambitious, several approximations and assumptions pertaining to this framework have proven quite viable. The most commonly applied methodology within density functional theory, the Kohn–Sham method, is actually quite similar to the Hartree–Fock method, which we addressed in relation to Exercise 4.4.4. However, contrary to the Hartree-Fock method, it does have the ability to include correlation energy in a single Slater determinant framework.

Yet another fundamental topic we have evaded is how quantum physics affects *statistical physics*. In Chapter 4 we saw that the spin of quantum particles cannot be disregarded – even when there is no explicit spin dependence in the Hamiltonian. The book-keeping must abide by the Pauli principle. Correspondingly, it is a matter of importance whether the particles we wish to describe by statistical means are fermions or bosons – particularly at low temperatures.

We have consistently assumed that we are well within the jurisdiction of quantum physical law. Yet we do know that the old Newtonian mechanics works very well for large objects – objects that are not microscopic. So where do Newton's laws cease to apply and quantum physics start to take over? This question, which has been addressed in Ref. [34], for instance, is far from trivial. We have learned, in Exercise 2.6.2, that the Schrödinger equation is consistent with Newton's second law if we interpret classical position and momentum as quantum physical expectation values. This is also in line with Bohr's *correspondence principle*, which claims that quantum physics reproduces the behaviour as dictated by classical laws in the case of very large quantum numbers. On the other hand, it is also a fact that several *semi-classical* approaches, methods in which parts of a quantum system are described by classical, Newtonian means, have proven able to describe quantum phenomena quite well. So it is fair to say that this borderline between classical physics and quantum physics is rather blurry.

For any of the topics we have addressed, you will have no trouble finding relevant and specialized literature. Nor will you have any trouble finding other introductions to quantum physics with a slightly different approach to the matter – typically a more analytical one. In this regard, it is worth mentioning John von Neumann's classic *Mathematical Foundations of Quantum Mechanics* from 1932, in which many key concepts were formalized (see Ref. [40] for a relatively new edition).

For every field where quantum physics applies, be it solid state physics, quantum chemistry, molecular and atomic physics, nuclear physics or particle physics, you will find extensive textbooks on the subject, books that may introduce you to the rich flora of phenomenology and methodology pertaining to each of them. To mention a few examples: Szabo and Ostlund's classic book *Modern Quantum Chemistry* is, albeit far from new, still a relevant introduction to many of the techniques used in the field [37]. A more extensive textbook in the field is provided by Helgaker, Jørgensen and

Olsen [20]. Bransden and Joachain have written an accessible introduction to molecular and atomic physics [11]. Within the field of solid state physics, the introduction written by Ashcoft and Mermin[2] [5], along with the one written by written by Kittel [24], are often applied. When it comes to *The Theory of Open Quantum Systems*, the book titled accordingly and written by Breuer and Petruccione is a classic [12], as is Nielsen and Chuang's book *Quantum Computation and Quantum Information* within its field [31]. Introductions to the world of nuclei and particles are provided, for instance, by Povh, Rith, Scholz, Zetsche and Rodejohann [32] and by Martin and Shaw [27]. In regard to the topic of the previous chapter, Moiseyev's book *Non-Hermitian Quantum Mechanics* is the go-to reference [30].

Needless to say, this listing far from exhaustive – by anyone's standards.

Some of these books are beyond the undergraduate level. Also for graduate students, while rewarding, working your way through a thorough textbook can be tedious. And sometimes a profound and technical familiarity is not essential either. Good popular books may often provide an interesting and informative introduction to the topic in question. In this spirit, we can recommend Bernhardt's book on quantum computing [10]. Zee has written an interesting story about how the notion of *symmetry* has played and continues to play a crucial role in the search for new insights within modern physics [41]. Hossenfelder, on the other hand, provides examples on how the same notion can also lead us astray [22].

Feynman wrote a popular introduction to the field of *quantum electrodynamics* [16]. Feynman was himself a very significant contributor to this field, which also set the ground for other quantum field theories.

Despite the fact that the first versions were written in the 1940s, George Gamow's books about *Mr Tompkins* and his encounters with modern physics [19] continue to educate and entertain readers, both young and old.

Much has been said and written about the meaning of quantum physics in general and the measurement problem in particular. One notable example is the book *Beyond Measure: Modern Physics, Philosophy and the Meaning of Quantum Physics*, by Bagott [6]. The book also gives an interesting account of the historical development of quantum physics.

Regardless of your future quantum path, we hope that this introduction has spurred your curiosity, and that this encounter with the beautiful, queer world of quantum physics has been a pleasant one.

[2] Yes, it *is* the same Mermin as the one we quoted earlier.

References

[1] E. Schneider. To Understand the Fourier Transform, Start from Quantum Mechanics, www.youtube.com/watch?v=W8QZ-yxebFA.

[2] Hitachi website, hitachi.com/rd/research/materials/quantum/doubleslit/.

[3] M. T. Allen, J. Martin, and A. Yacoby. Gate-defined quantum confinement in suspended bilayer graphene. *Nature Communications*, **3**(1):934, 2012.

[4] P. W Anderson. More is different: broken symmetry and the nature of the hierarchical structure of science. *Science*, **177**(4047):393–396, 1972.

[5] N. W. Ashcroft and N. D. Mermin. *Solid State Physics*. Holt, Rinehart and Winston, 1976.

[6] J. E. Baggott. *Beyond Measure: Modern Physics, Philosophy, and the Meaning of Quantum Theory*. Oxford University Press, 2004.

[7] E. Balslev and J. M. Combes. Spectral properties of many-body Schrödinger operators with dilatation-analytic interactions. *Communications in Mathematical Physics*, **22**(4):280–294, 1971.

[8] C. H. Bennett and G. Brassard. Quantum cryptography: public key distribution and coin tossing. arXiv preprint arXiv:2003.06557, 2020.

[9] C. H. Bennett and S. J. Wiesner. Communication via one- and two-particle operators on Einstein–Podolsky–Rosen states. *Physical Review Letters*, **69**:2881–2884, 1992.

[10] C. Bernhardt. *Quantum Computing for Everyone*. MIT Press, 2020.

[11] B. H. Bransden and C. J. Joachain. *Physics of Atoms and Molecules*. Prentice Hall, 2003.

[12] H. P. Breuer and F. Petruccione. *The Theory of Open Quantum Systems*. Oxford University Press, 2002.

[13] D. Bruton. Approximate RGB values for visible wavelengths. physics.sfasu.edu/astro/color/spectra.html.

[14] D. Chruściński and S. Pascazio. A brief history of the GKLS equation. *Open Systems & Information Dynamics*, **24**(03):1740001, 2017.

[15] R. P. Feynman. Simulating physics with computers. *International Journal of Theoretical Physics*, **21**(6):467–488, 1982.

[16] R. P. Feynman and A. Zee. *QED: The Strange Theory of Light and Matter*. Alix G. Mautner Memorial Lectures. Princeton University Press, 2006.

[17] A. B. Finnila, M. A. Gomez, C. Sebenik, C. Stenson, and J. D. Doll. Quantum annealing: a new method for minimizing multidimensional functions. *Chemical Physics Letters*, **219**(5):343–348, 1994.

[18] B. Friedrich and D. Herschbach. Stern and Gerlach: how a bad cigar helped reorient atomic physics. *Physics Today*, **56**(12):53–59, 2003.

[19] G. Gamow and R. Penrose. *Mr Tompkins in Paperback*. Canto. Cambridge University Press, 1993.

[20] T. Helgaker, P. Jørgensen, and J. Olsen. *Molecular Electronic-Structure Theory*. John Wiley & Sons, 2013.

[21] M. Hjorth-Jensen. The carbon challenge. *Physics*, **4**:38, 2011.

[22] S. Hossenfelder. *Lost in Math: How Beauty Leads Physics Astray*. Hachette UK, 2018.

[23] S. Jordan. The Quantum Algorithm Zoo, quantumalgorithmzoo.org/.

[24] C. Kittel and P. McEuen. *Introduction to Solid State Physics*. John Wiley & Sons, 8th edition, reprint edition, 2015.

[25] A. Kramida, Yu. Ralchenko, J. Reader, and NIST ASD Team. NIST Atomic Spectra Database (ver. 5.10), [Online]. Available at nist.gov/pml/atomic-spectra-database. National Institute of Standards and Technology, 2022.

[26] Y. I. Manin. Mathematics as metaphor. In *Proceedings of the International Congress of Mathematicians. Kyoto*, 1990.

[27] B. R. Martin and G. Shaw. *Nuclear and Particle Physics: An Introduction*. John Wiley & Sons, 2019.

[28] P. G. Merli, G. F. Missiroli, and G. Pozzi. On the statistical aspect of electron interference phenomena. *American Journal of Physics*, **44**(3):306–307, 1976.

[29] D. Mermin. What's wrong with this pillow? *Physics Today*, **42**(4):9, 1989.

[30] N. Moiseyev. *Non-Hermitian Quantum Mechanics*. Cambridge University Press, 2011.

[31] M. A. Nielsen and I. L. Chuang. *Quantum Computation and Quantum Information: 10th Anniversary Edition*. Cambridge University Press, 2010.

[32] B. Povh, K. Rith, C. Scholz, F. Zetsche, and W. Rodejohann. *Particles and Nuclei: An Introduction to the Physical Concepts*. Springer, 2015.

[33] A. Robinson. *The Last Man Who Knew Everything*. Oneworld Publications, 2006.

[34] M. A. Schlosshauer. *Decoherence and the Quantum to Classical Transition*. Springer Science & Business Media, 2007.

[35] B. Simon. Quadratic form techniques and the Balslev-Combes theorem. *Communications in Mathematical Physics*, **27**(1):1–9, 1972.

[36] J. Steeds, P. G. Merli, G. Pozzi, G. F. Missiroli, and A. Tonomura. The double-slit experiment with single electrons. *Physics World*, **16**(5):20, 2003.

[37] A. Szabo and N. S. Ostlund. *Modern Quantum Chemistry: Introduction to Advanced Electronic Structure Theory*. Dover Publications, 1996.

[38] M. Tokman, N. Eklöw, P. Glans, E. Lindroth, R. Schuch, G. Gwinner, D. Schwalm, A. Wolf, A. Hoffknecht, A. Müller, and S. Schippers. Dielectronic recombination resonances in F^{6+}. *Physical Review A*, **66**:012703, 2002.

[39] A. Tonomura, J. Endo, T. Matsuda, T. Kawasaki, and H. Ezawa. Demonstration of single-electron buildup of an interference pattern. *American Journal of Physics*, **57**(2):117–120, 1989.

[40] J. von Neumann. *Mathematical Foundations of Quantum Mechanics: New edition*. Princeton University Press, 2018.

[41] A. Zee. *Fearful Symmetry: The Search for Beauty in Modern Physics*. Collier Books, 1989.

Figure Credits

Fig. 1.1. Source: Dorling Kindersley RF/Getty Images.

Fig. 1.2. The Master and Fellows of Trinity College, Cambridge. Reference:Cambridge, Trinity College, Add.P.174(ii).

Fig. 1.3. Courtesy of the Niels Bohr Archive, Copenhagen.

Fig. 2.3. With the permission of Hitachi: hitachi.com/rd/research/materials/quantum/doubleslit/.

Fig. 2.5. Source: Professor Frank Trixler.

Fig. 2.6. Source: Jonas Dixon Østhassel.

Fig. 3.6. Facsimile of Mendeleev's 1869 periodic table of the elements. Public domain.

Fig. 3.7. Source: Unknown photographer, courtesy of the Museum of University History, University of Oslo.

Fig. 4.1. Source: Professor Horst Schmidt-Böcking.

Fig. 4.2. Courtesy of the Niels Bohr Archive, Copenhagen.

Fig. 4.3. Photograph by Paul Ehrenfest, courtesy of the Niels Bohr Archive, Copenhagen.

Fig. 4.4. Illustration by Professor Reiner Blatt, University of Innsbruck, used with permission.

Fig. 5.1. Source: M. T. Allen, J. Martin, and A. Yacoby. Gate-defined quantum confinement in suspended bilayer graphene. *Nature Communications*, **3**(1):934, 2012. With permission from Nature Customer Service Centre GmbH.

Fig. 6.2. Source: Professor Michael Schmid, TU Wien; adapted from the IAP/TU Wien STM Gallery / CC BY-SA 2.0.

Fig. 6.5. Fraunhofer lines. Public domain.

Fig. 6.7. Source: Ptrump16 / CC BY-SA 4.0.

Fig. 6.9. Source: CERN, used with permission.

Fig. 6.12. Media courtesy of D-Wave.

Fig. 7.1. Scriberia/Institute of Physics.

Fig. 7.2. From Dariusz Chruściński and Saverio Pascazio. A brief history of the GKLS equation. *Open Systems & Information Dynamics*, **24**(03):1740001, 2017. Used with the permission of Springer Nature BV, conveyed through Copyright Clearance Center, Inc.

Fig. 7.4. Portrait of Emmy Noether, around 1900. Public domain.

Fig. 7.5. Figure reprinted with permission, from M. Tokman, N. Eklöw, P. Glans, E. Lindroth, R. Schuch, G. Gwinner, D. Schwalm, A. Wolf, A. Hoffknecht, A. Müller, and S. Schippers, *Physical Review A* **66**, 012703 (2002). Copyright (2002) by the American Physical Society.

Fig. 8.5. Reproduced from 'Lise Meitner and Otto Hahn', Churchill Archives Centre, MTNR 8/4/1.

Fig. 9.1. Scriberia/Institute of Physics.

Index

absorption spectrum, 109
adiabatic theorem, 98, 99, 126
amplitude damping, 145, 147
Anderson, Philip W., 57
Ångström, Anders Jonas, 106
angular momentum, 41
annealing
  quantum, 128
  simulated, 128
annihilation, 136, 168
anti-commutator, 67, 133
anti-particle, 136
Aspect, Alain, 121
atom, 3
  Bohr atom, 4, 40
  helium, 70, 108
  hydrogen, 14
  one-dimensional model, 88
atomic units, 14
Auger–Meitner effect, 162
autoionization, 162
avoided crossing, 99, 126

b-spline, 91
Balmer series, 108
Balmer, Johann, 108
barrier
  double, 30, 155, 157
  general, 102
  linear, 104
  rectangular, 103
  smooth, 27
BB84, 123
Bell states, 62, 66, 120, 141
Bell, John Stewart, 66, 120, 121
Benett, Charles, 125
bipartite system, 140
black body radiation, 2
Bloch sphere, 116
Bloch's theorem, 44
Bloch, Felix, 45
Bohm, David, 166
Bohr formula, 40
Bohr radius, 14
Bohr, Niels, 4, 8, 40, 65, 166
Born interpretation, 7
  for momentum, 93
Born, Max, 7, 52, 93
Bose, Satyendra Nath, 59
boson, 59, 63
bound state, 37, 40
boundary conditions
  absorbing, 152
  Dirichlet, 24, 41
  outgoing, 103, 157
  periodic, 24, 45
bra, 13
bra-ket, 13
Brillouin, Léon, 104
Broglie, Louis de, 3, 34, 166

chemical shielding, 113
Clauser, John F., 121
coherent state, 48
commutator, 34, 67, 139
  fundamental, 34
complex absorbing potential, 152, 153
configuration interaction, 70
continuum, 39, 49, 95, 98, 149, 160, 162
  double, 163
  pseudo-continuum, 49, 95
correlation energy, 76
correspondence principle, 169
Coulomb interaction, 17
Coulomb potential, 14
coupled cluster, 77
Crank–Nicolson method, 92
creation, 136, 168
Curie, Marie, 5
curse of dimensionality, 6, 70, 77, 114
  blessing, 114

Davisson, Clinton, 26, 60
Davisson–Germer experiment, 60
decoherence, 122, 126, 144
degeneracy, 36
density functional theory, 168
density matrix, 138
  mixed state, 140
  positivity, 143
  pure state, 139
  reduced, 140, 141
  steady state, 146, 147
detuning, 83, 84, 112
dielectronic recombination, 150
diffusion equation, 51
dipole approximation, 88
Dirac equation
  negative energy solutions, 136
  non-relativistic limit, 137
  time-dependent, 132
    with electromagnetic field, 133
  time-independent, 134
Dirac notation, 13, 35, 115, 138
Dirac, Paul, 132
Dirichlet, Peter Gustav Lejeune, 24
discretize, 6, 17
dispersion, 4
dispersion relation, 45
displacement operator, 44
dissipation, 149, 151
double-slit experiment
  with electrons, 4, 26, 33
  with light, 2

Ehrenfest's theorem, 36, 169
Ehrenfest, Paul, 36, 65
eigenvalue equation, 32
  energy, 31, 37
    time-dependent, 97
  spin, 61
Einstein, Albert, 1, 3, 8, 65
electric field, 87
electron, 3
  charge, 14
  conduction, 104
  g-factor, 67, 137
  spin, 59
elementary charge, 14
emission spectrum, 109
Endo, J., 26
energy bands, 45
entanglement, 63, 120
Euler's method, 96
exchange symmetry, 58
excited state, 49
expectation value
  conservation, 35
  general, 12, 32
  kinetic energy, 21
  momentum, 10
  position, 8
Ezawa, H., 26

Fermi golden rule, 148
Fermi, Enrico, 59, 148
fermion, 59
Feynman, Richard, 6, 114, 165, 170
field theory, 136, 168
fifth Solvay conference, 5
finite difference
  double derivative
    five-point rule, 19
    three-point rule, 12, 19
  midpoint rule, 11
Fletcher, Harvey, 42
Fock, Vladimir, 76
Fourier transform, 19, 45, 93, 128
  fast, 20, 93
Fraunhofer, Joseph von, 110

Gaussian, 22, 23, 91
  potential, 43, 162
  wave function, 23
Gerlach, Walther, 57, 58, 60
Germer, Lester, 26, 60
GKLS equation, 142
Glauber state, 48
Glauber, Roy J., 48
Gordon, Walter, 132
Gorini, Vittorio, 142
Goudsmit, Samuel, 61
gradient descent, 53, 55, 72, 128
ground state, 49, 72, 89

Hadamard product, 56, 96
Hahn, Otto, 163
Hamilton, William Rowan, 15, 16
Hamiltonian, 15–17, 37
  Dirac, 132
    component form, 135
  laser interaction, 87
  non-Hermitian, 149, 152
  rotating wave, 83
  spin–spin interaction, 84
  time-dependent, 78
  two spin-1/2 particles, 85, 119
  two-state
    dynamic, 81
    static, 80, 81
  with electromagnetic field, 16
harmonic oscillator, 42, 99
Hartree, Douglas, 76
Hartree–Fock method, 74
  direct potential, 75, 76

exchange potential, 76
Hartree potential, 76
Heisenberg, Werner, 4, 101, 166
Hermicity, 13, 21, 46, 144
Hermitian adjoint, 13, 18, 35
Hermitian matrix, 22
Hoyle, Sir Fred, 164
Hylleraas, Egil, 51, 76

identity operator, 21, 85, 92, 123, 144
imaginary time, 50, 52, 96
inner product, 12, 61
interference, 2, 24–26, 94
ion trap, 66
ionization, 90, 93
Ionization threshold, 163
ionization threshold, 163
Ising model, 129

jump operator, 145

Kawasaki, T., 26
ket, 13
kinetic energy operator, 11, 19
three-dimensional, 16
Klein, Oskar, 132
Klein–Gordon equation, 131
Kohn–Sham method, 169
Kossakowski, Adrzej, 142
Kramers, Hendrik Anthony, 104
Kronecker delta function, 46
Kronecker product, 85
Kronig, Ralph, 61

laser, 78, 82, 88, 89
Lindblad equation, 142
Lindblad, Göran, 142
Lindbladian, 151
line spectrum, 108

magnetic moment, 80
magnetic resonance imaging, 110
Magnus propagator, 87
Magnus, Wilhelm, 87
Manin, Yuri, 6
master equation, 138, 142
Matsuda, T., 26
Maxwell, James Clerk, 2
measurement, 7, 8, 11, 32, 41
probability, 33
Meitner, Lise, 162
Mendeleev, Dimitri, 49, 50
Merli, P. G., 26
Mermin, David, 166
metrology, 101
Michelson–Morley experiment, 61
Millikan, Robert A., 42
Missiroli, G. F., 26
mixed state, 140
momentum distribution, 93
momentum operator, 10
three-dimensional, 16

Neumann, John von, 98, 139, 169
Newton's second law, 1, 28, 35
Newton, Isaac, 1
Noether's theorem, 147, 151
Noether, Emmy, 147
non-crossing rule, 98
non-locality, 65, 157

oil drop experiment, 42
operator exponential, 21
optical cycle, 88, 89
optimization, 100, 126
ordinary differential equation, 29
orthonormal, 46, 91
overlap matrix, 91

partial differential equation, 16, 90
partial trace, 139
Pauli matrices, 67, 132, 133
characteristics, 67
eigenvalues, 68
Pauli principle, 45, 59, 63, 169
Pauli rotation, 81, 118
Pauli, Wolfgang, 59, 61, 67, 166
periodic potentials, 43
periodic table, 49, 76
perturbation theory, 168
photoelectric effect, 3
photoionization, 88, 153
photon, 3, 17, 40, 168
polarization, 115
spin, 59, 109
piezoelectricity, 107
pilot waves, 166
Planck constant, 3
reduced, 10, 11
Planck, Max, 2
position operator, 10
Pozzi, G., 26
probability current, 102
probability rate, 148
propagator, 21, 86, 117
proton
g-factor, 110
mass, 110
spin, 110
pure state, 138, 139
purity, 140

quantization
angular momentum, 42
projection, 59

charge, 42
electromagnetic field, 168
energy, 38, 39
harmonic oscillator, 42
spin, 57, 58
projection, 59
quantum advantage, 122
quantum computing
adiabatic, 97, 126
algorithms, 122
annealing, 126, 129
gate-based, 117
quantum dot, 78, 79
quantum foundations, 166
quantum gates, 117
CNOT, 119, 120
fidelity, 119
Hadamard, 117, 120
NOT, 117, 120
SWAP, 119
quantum guitar, 33, 41
quantum interpretation
shut up and calculate, 166
Born interpretation, 7, 93
Copenhagen interpretation, 8, 166
de Broglie–Bohm interpretation, 166
quantum key distribution, 123
quantum parallelism, 122
quantum sheep, 30
quantum statistics, 169
quantum strangeness, 166
qubit, 86, 115, 116

Rabi frequency, 84, 111
generalized, 84, 111
Rabi, Isodor Isaac, 84
reflection probability, 28, 103, 155
relativity
general, 1
special, 61, 130
resonance, 112
doubly excited state, 162
exponential decay, 160
lifetime, 159, 160
nuclear magnetic resonance, 110
shape resonance, 156
width, 159
rotating wave approximation, 82, 111
Runge–Kutta method, 29, 90

scanning tunnelling microscope, 31, 105
scattering, 27, 102, 154, 155
Schrödinger equation, 5
time-dependent, 16, 37, 81
in adiabatic basis, 98
matrix form, 91
spectral representation, 91
time-independent, 31
Hartree, 75
non-linear, 75
periodic potential, 44
Schrödinger, Erwin, 4, 16, 34, 132
selection rule, 109
self-consistent field, 74
singular value decomposition, 144
Slater determinant, 69
Slater permanent, 69
Slater, John C., 69
spectral theorem, 22, 46, 90
spectroscopy, 110
speed of light, 65, 108, 135
spherical harmonics, 42
spin, 57, 74
dynamics, 78, 81, 84, 141
flip, 81, 82, 110
magnetic interaction, 67, 79
singlet, 63, 70
spin–spin interaction, 84, 85
triplet, 63, 70
spinor, 62, 134
split operator, 87, 100, 127
standard deviation, 12
for eigenstates, 32
general, 12
momentum, 12
position, 8
Stern, Otto, 57, 58
Stern–Gerlach experiment, 58, 61
Sudarshan, George, 142
superdense coding, 123

tensor product, 71, 85
thermal state, 142
time dilation, 131
Tonomura, A., 26
translation operator, 44
transmission probability, 28, 103, 155
trapezoidal rule, 9, 86
tunnelling, 29, 30, 102, 128

Uhlenbeck, George, 61
uncertainty relation, 4, 7, 12, 25
unitary transform, 82, 117

variational principle, 50, 52, 53, 56, 70, 72, 73
vector potential, 17, 134, 168
von Neumann entropy, 140
von Neumann equation, 139

wave function, 5
collapse, 7, 9, 33, 68, 122
Dirac, 134
expansion in eigenstates, 46, 62, 98
Gaussian, 25, 30

Hartree–Fock, 76
normalization, 7, 18, 47, 62, 115
product state, 61, 62
spin part, 61
stationary solution, 31, 102
steady current, 101
two particles, 95
with spin and spatial dependence, 61, 134

Wentzel, Gregor, 104
Wigner, Eugene, 98
WKB approximation, 104
work function, 106

Young, Thomas, 2, 25

Zeilinger, Anton, 121